新技术技能人才培养系列教程

大数据核心技术系列

北京课工场教育科技有限公司 出品

Hadoop 应用开发基础

刘雯 王文兵 / 主编

倪天伟 郭迎慧 李贤志 / 副主编

人 民 邮 电 出 版 社

北 京

图书在版编目（CIP）数据

Hadoop应用开发基础 / 刘雯，王文兵主编. -- 北京：人民邮电出版社，2019.1(2021.1重印)
新技术技能人才培养系列教程
ISBN 978-7-115-49813-7

Ⅰ. ①H… Ⅱ. ①刘… ②王… Ⅲ. ①数据处理软件－程序设计－教材 Ⅳ. ①TP274

中国版本图书馆CIP数据核字(2018)第241029号

内 容 提 要

Hadoop 是一个分布式系统的基础架构，支持对大量数据进行分布式处理，能以高效、可靠的方式完成数据处理。本书围绕 Hadoop 生态圈技术进行讲解，主要包括 Hadoop 环境配置、Hadoop 分布式文件系统（HDFS）、Hadoop 分布式计算框架 MapReduce、Hadoop 资源调度框架 YARN 与 Hadoop 新特性、Hadoop 分布式数据库 HBase、Oozie 工作流调度系统等内容。

本书以 Linux 操作系统为平台，紧密结合实际应用，贯穿了大量实践案例。另外，本书配以多元的学习资源和平台服务，包括参考教案、案例素材下载、学习交流社区等，为读者提供全方位的学习体验。通过系统地学习本书内容和操作实践，读者可以掌握大数据相关技能。

本书适合作为计算机专业大数据等相关课程的教材使用，也适合具有一定 Linux、Java 开发经验且想从事大数据开发工作的人员自学使用，还适合作为大数据分析与运维人员的参考用书。

◆ 主　　编　刘　雯　王文兵
副 主 编　倪天伟　郭迎慧　李贤志
责任编辑　祝智敏
责任印制　马振武

◆ 人民邮电出版社出版发行　　北京市丰台区成寿寺路 11 号
邮编　100164　　电子邮件　315@ptpress.com.cn
网址　http://www.ptpress.com.cn
固安县铭成印刷有限公司印刷

◆ 开本：787×1092　1/16
印张：15.75　　　　2019 年 1 月第 1 版
字数：335 千字　　　2021 年 1 月河北第 4 次印刷

定价：46.80 元

读者服务热线：(010)81055256　印装质量热线：(010)81055316
反盗版热线：(010)81055315
广告经营许可证：京东市监广登字20170147号

大数据核心技术系列

编　委　会

序　言

丛书设计

大数据已经悄无声息地改变了我们的生活和工作方式，精准广告投放、实时路况拥堵预测已很普遍，在一些领域，人工智能比我们更加聪明、高效，未来的个性化医疗、教育将会真正实现，大数据迎来前所未有的机遇。Google 公司 2003 年开始陆续发表的关于 GFS、MapReduce 和 BigTable 的三篇技术论文，成为大数据发展的重要基石。十几年来大数据技术从概念走向应用，形成了以 Hadoop 为代表的一整套大数据技术。时至今日，大数据技术仍在快速发展，基础框架、分析技术和应用系统都在不断演变和完善，并不断地涌现出大量新技术，成为大数据采集、存储、处理、分析、可视化呈现的有效手段。企业需要利用大数据更加贴近用户、加强业务中的薄弱环节、规范生产架构和策略。对数家企业的调查显示，大数据工程师应该掌握的技能包括：Hadoop、HDFS、MapReduce、Hive、HBase、ZooKeeper、YARN、Sqoop、Spark、Spark Streaming、Scala、Kafka、Confluent、Flume、Redis、ETL、Flink/Streaming、Linux、Shell、Python、Java、MySQL、MongoDB、NoSQL、Cassandra、Spark MLib、Pandas、Numpy、Oozie、ElasticSearch、Storm 等，作为一名大数据领域的初学者，在短时间内很难系统地掌握以上全部技能点。“大数据核心技术系列”丛书根据企业人才实际需求，参考以往学习难度曲线，选取“Hadoop+Spark+Python”技术集作为核心学习路径，旨在为读者提供一站式、实战型大数据开发学习指导，帮助读者踏上由开发入门到实战的大数据开发之旅！

“大数据核心技术系列”以 Hadoop、Spark、Python 三个技术为核心，根据它们各自不同的特点，解决大数据中离线批处理和实时计算两种主要场景的应用。以 Hadoop 为核心完成大数据分布式存储与离线计算；使用 Hadoop 生态圈中的日志收集、任务调度、消息队列、数据仓库、可视化 UI 等子系统完成大数据应用系统架构设计；以 Spark Streaming、Storm 替换 Hadoop 的 MapReduce 以实现大数据的实时计算；使用 Python 完成数据采集与分析；使用 Scala 实现交互式查询分析与 Spark 应用开发。书中结合大量项目案例完成大数据处理业务场景的实战。

在夯实大数据领域技术基础的前提下，“大数据核心技术系列”丛书结合当下 Python 语言在数据科学领域的活跃表现以及占有量日益扩大的现状，加强了对 Python 语言基础、Scrapy 爬虫框架、Python 数据分析与展示等相关技术的讲解，为读者将来在大数据科学领域的进一步提升打下坚实的基础。

丛书特点

1．以企业需求为设计导向

满足企业对人才的技能需求是本系列丛书的核心设计原则，课工场大数据开发教研团队通过对数百位 BAT 一线技术专家进行访谈、对上千家企业人力资源情况进行调研、对上万个企业招聘岗位进行需求分析，实现对技术的准确定位，达到课程与企业需求的高契合度。

2．以任务驱动为讲解方式

丛书中的知识点和技能点均由任务驱动，读者在学习知识时不仅可以知其然，而且可以知其所以然，帮助读者融会贯通、举一反三。

3．以实战项目来提升技术

本丛书均设置项目实战环节，以综合运用书中的知识点帮助读者提升项目开发能力。每个实战项目都设有相应的项目思路指导、重难点讲解、实现步骤总结和知识点梳理。

4．以“互联网+”实现终身学习

本丛书可配合课工场 App 进行二维码扫描，来观看配套视频的理论讲解和案例操作，同时课工场在线开辟教材配套版块，提供案例代码及案例素材下载。此外，课工场还为读者提供了体系化的学习路径、丰富的在线学习资源和活跃的学习社区，方便读者随时学习。

读者对象

1．大中专院校的学生

2．编程爱好者

3．初中级程序开发人员

4．相关培训机构的老师和学员

读者服务

学习本丛书过程中如遇到疑难问题，读者可以访问课工场在线，也可以发送邮件到 ke@kgc.cn，我们的客服专员将竭诚为您服务。

感谢您阅读本丛书，希望本丛书能成为您大数据开发之旅的好伙伴！

“大数据核心技术系列”丛书编委会

前　言

大数据技术让我们以前所未有的方式对海量数据进行分析，从中获得有巨大价值的产品和服务，最终形成变革之力。如何从零基础学习大数据平台 Hadoop 的应用开发技术，并运用相关技术解决一些实际的业务需求，正是本书的编写目的。全书共 10 章，各章主要内容如下。

第 1 章是对 Hadoop 的总体概述，包括大数据的基本概念、Hadoop 生态圈、Hadoop 与大数据的关系，以及 Hadoop 安装部署的详细步骤。

第 2 章是对 HDFS 的介绍，主要包括 HDFS 的体系结构、Shell 操作以及使用 Java API 访问 HDFS 系统。

第 3 章是对 MapReduce 分布式计算框架的讲解，包括 MapReduce 的编程模型、编写和运行 MapReduce 程序，同时还配以多个经典的 MapReduce 应用案例。

第 4 章是对 Hadoop 新的资源调度框架 YARN 以及 Hadoop 新特性的讲解，并深度分析了如何实现 Hadoop 高可用集群及高可用的实现原理。

第 5 章是对 ZooKeeper 分布式高可靠协调服务的讲解，主要介绍 ZooKeeper 的架构设计以及数据模型，解析如何掌握 ZooKeeper 单机环境的搭建，并能利用 ZooKeeper 实现分布式系统服务器上下线的动态感知。

第 6 章是对 HBase 数据库的基础讲解，介绍 HBase 的体系架构及数据模型，分析如何实现 HBase 的伪分布式环境搭建。

第 7 章是对 HBase 操作的实践讲解，详细介绍 HBase 的 DDL、DML 的 Shell 操作以及如何用 Java API 实现对《王者荣耀》游戏玩家信息表的管理。

第 8 章是对 HBase 的高级知识扩展，深度解析 HBase 表空间管理和权限管理、HRegion 的切分原理、HBase 中的 Compaction 过程，提升读者对 HBase 的认知。

第 9 章是对 Oozie 调度框架的讲解，介绍 Oozie 的架构及执行流程，引导读者搭建 Oozie 环境，并在 Oozie 上进行作业的调度。

第 10 章的综合项目实训利用前面各章所学的 Hadoop 生态圈中的 HDFS、YARN、ZooKeeper、HBase 等知识，自主开发《王者荣耀》游戏英雄排行榜功能，通过理论与实践的结合来加强读者对知识的掌握和运用。

读者学习大数据技术，就要多动手练习，从而深入理解每个知识点，提高编程熟练度，培养分析问题和解决问题的能力，不断积累开发经验。同时，学习中读者还要通过交流来消除学习疑惑，分享学习经验，取长补短，共同进步。

本书提供了便捷的学习体验，读者可以通过扫描二维码下载各章提供的资源，包括素材、技能实训源码及本章作业参考答案等。

本书由课工场大数据开发教研团队编写，参与编写的还有刘雯、王文兵、倪天伟、郭迎慧、李贤志等院校老师。尽管编者在写作过程中力求准确、完善，但书中不足或疏漏之处仍在所难免，殷切希望广大读者批评指正！

智慧教材使用方法

扫一扫查看视频介绍

由课工场“大数据、云计算、全栈开发、互联网 UI 设计、互联网营销”等教研团队编写的系列教材，配合课工场 App 及在线平台的技术内容更新快、教学内容丰富、教学服务反馈及时等特点，结合二维码、在线社区、教材平台等多种信息化资源获取方式，形成独特的“互联网+”形态——智慧教材。

智慧教材为读者提供专业的学习路径规划和引导，读者还可体验在线视频学习指导，按如下步骤操作可以获取案例代码、作业素材及答案、项目源码、技术文档等教材配套资源。

1．下载并安装课工场 App。

（1）方式一：访问网址 www.ekgc.cn/app，根据手机系统选择对应课工场 App 安装，如图 1 所示。

图1　课工场App

（2）方式二：在手机应用商店中搜索“课工场”，下载并安装对应 App，如图 2、图 3 所示。

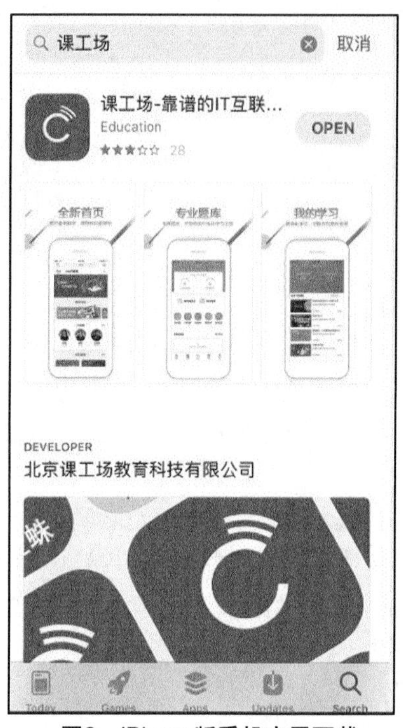

图2　iPhone版手机应用下载

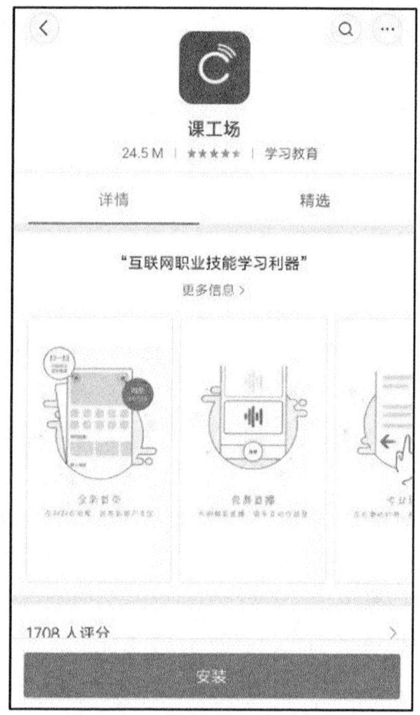

图3　Android版手机应用下载

2．登录课工场 App，注册个人账号，使用课工场 App 扫描书中二维码，获取教材配套资源，依照如图 4 至图 6 所示的步骤操作即可。

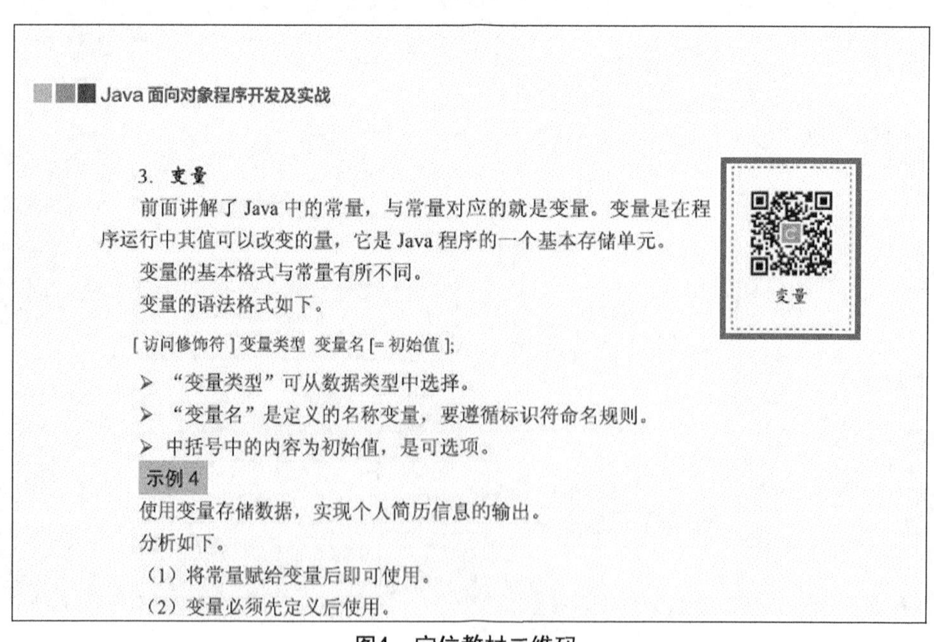

Java 面向对象程序开发及实战

3．变量

前面讲解了 Java 中的常量，与常量对应的就是变量。变量是在程序运行中其值可以改变的量，它是 Java 程序的一个基本存储单元。

变量的基本格式与常量有所不同。

变量的语法格式如下。

[访问修饰符] 变量类型　变量名 [= 初始值];

➢ “变量类型”可从数据类型中选择。
➢ “变量名”是定义的名称变量，要遵循标识符命名规则。
➢ 中括号中的内容为初始值，是可选项。

示例 4

使用变量存储数据，实现个人简历信息的输出。

分析如下。

（1）将常量赋给变量后即可使用。

（2）变量必须先定义后使用。

变量

图4　定位教材二维码

图5　使用课工场App“扫一扫”扫描二维码

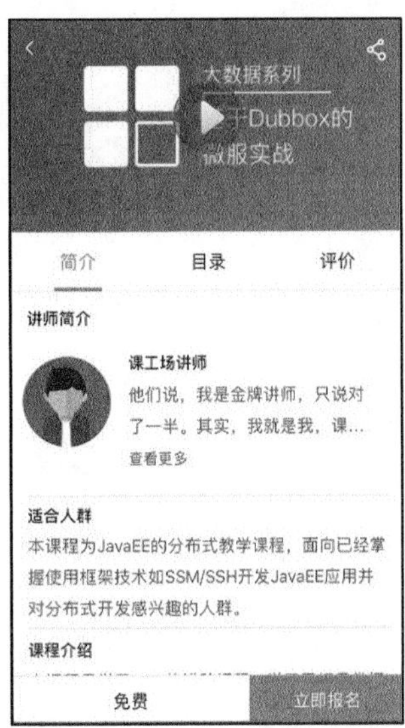

图6　使用课工场App免费观看教材配套视频

3．获取专属的定制化扩展资源。

（1）普通读者请访问 http://www.ekgc.cn/bbs 的“教材专区”版块，获取教材所需开发工具、教材中示例素材及代码、上机练习素材及源码、作业素材及参考答案、项目素材及参考答案等资源（注：图 7 所示网站会根据需求有所改版，下图仅供参考）。

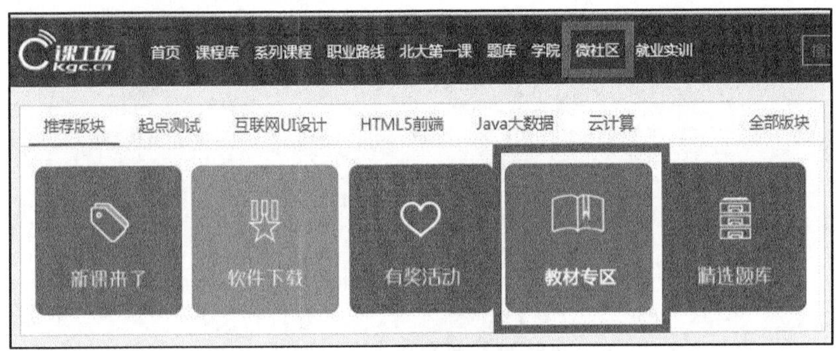

图7　从社区获取教材资源

（2）高校老师请添加高校服务 QQ：1934786863（如图 8 所示），获取教材所需开发工具、教材中示例素材及代码、上机练习素材及源码、作业素材及参考答案、项目素材及参考答案、教材配套及扩展 PPT、PPT 配套素材及代码、教材配套线上视频等资源。

图8　高校服务QQ

目 录

第1章 Hadoop入门

技能目标

- 了解大数据和 Hadoop 概念
- 掌握 Hadoop 架构及核心构成
- 了解 Hadoop 生态圈技术
- 能够搭建 Hadoop 平台
- 能够运行 Hadoop 程序

本章任务

任务 1　了解大数据现状

任务 2　了解 Hadoop 基础

任务 3　搭建移动通信业务的 Hadoop 处理平台

本章资源下载

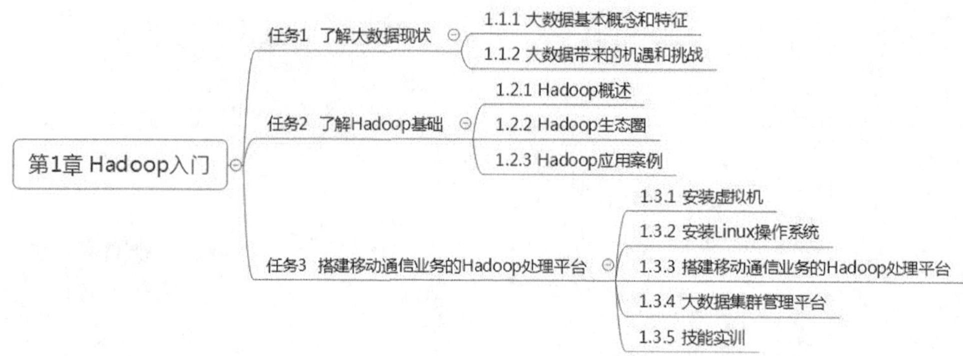

在当今大数据的时代背景下，Hadoop 作为大数据处理领域的分布式存储和计算框架，已经得到了众多国内外企业的青睐，并得到广泛使用。对于从事大数据工作的开发人员来说，掌握 Hadoop 技术是非常必要的。本章主要介绍目前大数据的现状和特征、Hadoop 框架的核心构成、Hadoop 生态圈技术及应用场景，同时介绍如何搭建 Hadoop 平台。

任务 1 了解大数据现状

【任务描述】

了解大数据的概念及特征，了解大数据带来的机遇、挑战及应对策略。

【关键步骤】

（1）了解大数据的概念及特征。

（2）了解大数据时代的机遇与挑战，以及如何应对。

1.1.1 大数据基本概念和特征

1. 什么是大数据

“大数据”作为当今最热门的 IT 行业词汇，在互联网时代变得越来越重要。究竟什么是大数据？大数据是指无法在一定时间内用常规软件工具对其内容进行抓取、管理和处理的数据集合。对于“大数据”（Big Data），研究机构 Gartner 给出这样的定义：大数据是需要新处理模式才能具有更强的决策力、洞察发现力和流程优化能力的海量、高增长率和多样化的信息资产。

2. 大数据特征

（1）4V + 1O 特征

大数据量（Volume）。采集、存储和计算的数据量大。大数据时代下，每时每刻都在产生着大量的数据，比如社交网络，交通等领域，每天都会产生很多的日志文件。大数据的起始计量单位至少是 PB 量级的。

类型繁多（Variety）。数据种类和来源多样化。数据的种类包括结构化、半结构化和非结构化数据，具体表现为网络日志、音频、视频、图片、地理位置信息等。多样化的数据对大数据处理技术提出了更高的要求。

价值密度低（Value）。随着现阶段物联网的广泛应用，接入到互联网的信息感知设备无处不在，产生了海量的数据，但数据价值密度较低，如何结合业务逻辑并通过强大的机器学习算法来挖掘数据价值，是大数据时代最需要解决的问题。

速度快、时效高（Velocity）。数据增长速度快、处理速度快，时效性要求高。在使用搜索引擎时，用户希望几分钟前的新闻能够被查到，个性化推荐算法尽可能要求实时完成推荐。这是大数据区别于传统数据的显著特征。

数据在线（Online）。数据是永远在线的，是随时能调用和计算的，这是大数据区别于传统数据最大的特点。现在提及大数据不仅仅是“大”，更重要的是数据变得在线了，这是互联网发展背景下赋予大数据的时代特征。

（2）固有特征

时效性。数据在某一时间段内具有对决策有价值的属性，也就是说，同一信息在不同的时间具有很大的性质上的差异，这个差异就是数据的时效性。信息的时效性决定了决策在哪些时间内有效。

不可变性。数据不会改变，也就是说，大数据的变化可看作是新产生的数据条目，而不是对现有条目的更新。

1.1.2 大数据带来的机遇和挑战

随着互联网和云计算的飞速发展，物联网和社交网络的日益普及，当前社会已进入大数据时代。大数据作为一个时代、一项技术、一个挑战、一种文化，对社会的发展带来了深刻的影响。党的十八届五中全会指出，要实施“国家大数据战略”。实施国家大数据战略，必须正确认识大数据，准确把握其带来的机遇，科学应对其带来的挑战，用大智慧实现大数据的价值。

1. 机遇

（1）大数据已经成为重要的战略资源。在当今社会，资源已经不仅仅指传统的矿产、石油等资源，大数据等信息资源也成为重要的战略资源。在各行各业，每分每秒都在产生数据，企业可以通过这些数据了解市场和用户需求，做出精准营销。大数据应用已经成为提高企业核心竞争力的关键因素，数据资产逐渐成为商业社会的核心竞争力。越来越多的企业开始重视大数据战略布局。

（2）大数据人才需求大幅增长。进入大数据时代以来，对大数据处理技术人才的需求呈现爆炸式的增长，企业提供了更多的岗位需求，为想要技术转型及转换行业的人员提供了一个不错的选择。

2. 挑战

大数据时代带来了机遇的同时，也带来了一定的挑战。

（1）对现有存储方式的挑战

传统的数据存储是将数据存储在数据库中，随着大数据时代的到来，传统的数据存储方式已经不能适应存储 PB 量级的数据。同时新产生的数据具有多样化的特点，一些非结构化的数据也不能采用传统的结构化数据系统存储。

（2）对现有企业的挑战

大数据环境下，企业提供业务服务的传统运营模式已经不具优势，必须逐步向数据服务转型。目前企业面临的最显著挑战就是数据的碎片化，在很多企业尤其是大型企业，数据常常散落在不同部门，导致企业内部的数据无法打通，大数据的价值难以挖掘。

（3）对技术的挑战

由于大数据具有时效性强的特点，数据的价值会随着时间的流逝而降低，这就要求对数据进行快速的处理，电商以及新闻资讯的推荐系统就是很好的案例。假如推荐的内容延迟性太高，对于订单的转化效果就会大大降低，实时推荐就是大数据技术对传统技术的挑战。

3．**应对策略**

如何应对大数据时代下的挑战呢？

（1）培养大数据专业人才。大数据技术有着很高的门槛，大数据建设的每一个环节都需要专业的人员完成，因此必须培养更多的掌握大数据技术的专业人才，这就需要高校和企业共同努力培养和挖掘。

（2）加快新技术的研发和创新。大力研发大数据新技术并重视其应用。

任务 2　了解 Hadoop 基础

【任务描述】

了解 Hadoop 概念及由来，掌握 Hadoop 的核心组件及组件功能，了解 Hadoop 生态圈技术的功能。

【关键步骤】

（1）认识 Hadoop。

（2）掌握 Hadoop 的核心组件及功能。

（3）了解 Hadoop 的生态圈技术及应用场景。

（4）了解大数据在目前行业中的应用案例。

1.2.1　Hadoop 概述

1．**什么是 Hadoop**

Hadoop 是由 Apache 软件基金会开发的一个可靠的、可扩展的分布式系统架构。架构中包含用于解决大数据存储的分布式文件系统（Hadoop Distributed File System，HDFS）、用于解决分布式计算的分布式计算框架 MapReduce 以及分布式资源管理系统 YARN。Apache Hadoop 软件库是一个框架，允许用户在不了解分布式系统底层细节的情况下，使用简单的编程模型开发分布式程序，并充分利用集群的分布式能力进行运算和存储。它的设计目的是从单一的服务器扩展到成千上万的机器，并将集群部署在多台机器中，每台机器提供本地计算和存储。Apache Hadoop 生态圈已成为目前处理海量数据的首选架构。

2. Hadoop 发展史

Apache Hadoop 起源于开源的网络搜索引擎 Apache Nutch，Nutch 是 Apache Lucene 项目的一部分。2002 年，Apache Lucene 的创始人 Doug Cutting 创建了 Hadoop。

2003—2004 年，Google 发表了 *The Google File System* 和 *MapReduce:Simplifed Data Processing on Large Cluster* 两篇论文，向全世界展示了 Google 分布式文件系统（GFS）和 MapReduce 框架。

2005 年年初，Nutch 开发人员在 Nutch 上实现了一个 MapReduce 算法，花费半年左右的时间完成 Nutch 主要算法的移植，并用 MapReduce 和 NDFS 来运行。

2006 年 2 月，开发人员将 NDFS 和 MapReduce 与 Nutch 分离，形成 Lucene 子项目，并命名为 Hadoop。Doug Cutting 几经周折加入 Yahoo 公司，并致力于 Hadoop 技术的进一步发展。

2008 年 1 月，Hadoop 成为 Apache 的顶级项目，同年 4 月，Hadoop 打破世界纪录，成为最快的 TB 量级数据排序系统。

2009 年 3 月，Cloudera 公司基于 Apache Hadoop 发布了 CDH 版本。

2011 年 12 月，Hadoop 发布 1.0.0 版本，标志着 Hadoop 已经初具生产规模。

2013 年，Hadoop 发布了 2.2.0 版本，Hadoop 进入到 2.x 时代。

2014 年，Hadoop2.x 更新速度加快，先后发布了 Hadoop2.3.0、Hadoop2.4.0、Hadoop2.5.0 和 Hadoop2.6.0，极大地完善了 YARN 框架和整个集群的功能。

2015 年，发布了 Hadoop2.7.0 版本。

2016 年，Hadoop 及其生态圈在各行各业落地并且得到广泛应用，同年，Hadoop 发布 Hadoop3.0-alpha 版本，标志着 Hadoop 进入 3.x 时代。

3. Hadoop VS RDBMS

在许多场景下，Hadoop 能够被视为 RDBMS（关系型数据库管理系统）的一种补充。两个系统之间的对比如表 1-1 所示。Hadoop 很适合那些需要分析（尤其是自主分析）整个数据集的问题，以批处理的方式进行，而 RDBMS 适合于点查询和更新。Hadoop 适合数据被一次写入和多次读取的应用，而 RDBMS 适合持续更新的数据集。

表 1-1　RDBMS 与 Hadoop 对比

	RDBMS	Hadoop
数据大小	GB	TB
更新	多次读写	一次写入、多次读取
速度	读数据速度快	写数据速度快
数据类型	结构化	结构化、半结构化、非结构化
适用场景	交互式联机分析处理 复杂事务处理	处理非结构化数据 海量数据存储和计算

4. Hadoop 核心构成

Hadoop 框架包括三个部分：分布式文件系统 HDFS、计算系统 MapReduce、资源管理系统 YARN。

（1）分布式文件系统 HDFS

HDFS 是谷歌 GFS 的克隆版，是对谷歌 2003 年 10 月发表的 GFS 论文的开源实现。作为大数据领域的数据存储，HDFS 的设计目标就是提供一个具有高可靠性、高容错性、高吞吐量以及能运行在通用硬件上的分布式文件存储系统。

HDFS 的设计思想是将数据文件以指定的大小切分成数据块，将数据块以多副本的方式存储在多台机器上。这样的设计使 HDFS 可以更方便地做数据负载均衡以及容错，而且数据文件的切分、数据负载均衡和容错这些功能对用户都是透明的，用户在使用的时候，可以把 HDFS 当作普通的本地文件系统使用。

（2）分布式计算框架 MapReduce

MapReduce 是 Hadoop 的核心计算框架，用于 PB 量级数据的并行计算。MapReduce 是一种简化应用程序开发的编程模型，模型中主要包括 Map（映射）和 Reduce（规约）两项核心操作。MapRecude 编程模型为应用开发者隐藏了系统层实现细节，允许用户不必关注并行计算底层实现，只需按照 MapReduce API 的编程模型即可实现相应业务逻辑的开发。

当启动一个 MapReduce 任务时，作业会将输入的数据集切分成若干独立的数据块，由 Map 端将数据映射成需要的键值对类型，然后对 Map 的输出进行排序，再把结果输入 Reduce 端；Reduce 端接收 Map 端传过来的键值对类型的数据，根据不同键分组，对每一组键相同的数据进行处理，得到新的键值对并输出，这就是 MapReduce 的核心思想。通常 MapRedude 任务的输入和输出都是使用 HDFS 进行存储，也就是说，MapReduce 处理数据的大部分场景都存储在 HDFS 上。

（3）资源管理系统 YARN

YARN（Yet Another Resource Negotiator，另一种资源协调者）是一种新的 Hadoop 资源管理器。在 Hadoop1.x 版本中还没有 YARN，它的出现解决了 Hadoop1.x 版本中 MapReduce 架构中 JobTracker 负载压力过大的问题，它将 JobTracker 的资源管理和作业调度拆分成两个独立的服务，分别为全局的资源管理器（ResourceManager）和每个应用程序特有的 ApplicationMaster。其中，ResourceManager 负责整个系统的资源管理和分配，而 ApplicationMaster 负责单个应用程序的管理。

YARN 是随着 Hadoop 的不断发展而催生的新框架，它的引入不仅解决了 JobTracker 负载压力大的问题，同时也解决了 Hadoop1.x 中只能运行 MapReduce 作业的限制。YARN 作为一个通用的资源管理系统，允许在其上运行各种不同类型的作业，比如 MapReduce、Spark、Tez 等。

YARN 的引入，为 Hadoop 集群在利用率、资源统一管理和数据共享等方面带来了极大的提升。

5. *为什么选择 Hadoop 作为大数据的解决方案*

企业在选择技术架构的时候，主要考量的方面包括源码的开放程度、社区的活跃程度、目前在业界的使用情况等。Hadoop 能被选作大数据的解决方案，原因有很多，本节主要列举以下四个原因。

（1）Hadoop 源代码开放。

（2）社区活跃、参与者众多。实际工作中遇到问题可以在社区得到很好的解决。

（3）Hadoop 发展到现在，已经得到企业界的广泛验证。

（4）Hadoop 生态圈可胜任分布式存储和计算的各个场景。

6. Hadoop *发行版本*

目前而言，Hadoop 的发行版本主要有三个。

（1）Apache Hadoop。这是最原始的版本，所有的发行版都是基于这个版本进行改进的，也被称为社区版 Hadoop。

（2）Cloudera CDH。是目前国内公司使用最多的。CDH 完全开源，比起 Apache Hadoop，在兼容性、安全性和稳定性上均有所增强。

（3）Hortonworks HDP。该版本的 Hadoop 是百分之百开源的，版本和社区版完全一致，它集成了开源监控方案 Ganglia 和 Negios。2018 年 10 月，两家大数据先驱 Cloudera 和 Hortonworks 宣布合并。

1.2.2　Hadoop 生态圈

1. *概述*

Hadoop 自出现以后，得到快速发展，大量与其相关的应用也被开发出来，共同服务于 Hadoop 工程。现在 Hadoop 已经成为一个庞大的架构体系，只要是与大数据相关的，都会出现 Hadoop 的身影。这些相关组件系统与 Hadoop 一起构成 Hadoop 生态圈，如图 1.1 所示。

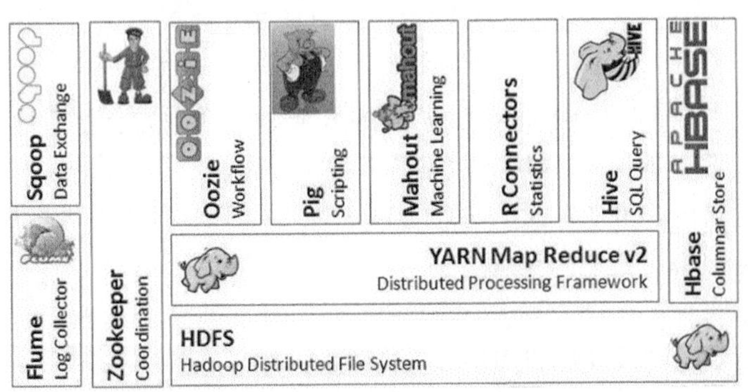

图1.1　Hadoop生态圈

2. Hadoop *生态圈技术*

（1）Hadoop 核心

Hadoop 的核心构成包括用于分布式存储的 HDFS、用于分布式计算的 MapReduce，以及用于分布式资源管理的 YARN。三个核心的具体功能及架构在后面的章节会详细讲解。

（2）数据查询分析

Hadoop 生态圈提供了方便用户使用的数据查询分析框架 Hive 和 Pig。下面分别对这两种框架进行介绍。

Hive 是建立在 Hadoop 之上的数据仓库基础框架，可以将结构化的数据文件映射为一张数据库表，并定义了一种类 SQL 语言（HQL），让不熟悉 MapReduce 的开发人员也能编写数据查询语句来对大数据进行分析统计操作。Hive 的出现极大地降低了大数据技术的学习门槛，同时提高了开发效率。

Pig 是一个基于 Hadoop 的大规模数据分析平台，它有一套叫作 Pig Latin 的类 SQL 语言，该语言的编译器会把类 SQL 的数据分析请求转换成一系列经过优化处理的 MapReduce 运算，处理的对象是 HDFS 上的文件。

（3）协调管理

在 Hadoop 生态圈中，使用 ZooKeeper 框架来解决分布式环境下的数据管理问题，比如统一命名、状态同步和配置同步等问题。Hadoop 的大多数组件都依赖于 ZooKeeper，比如 HBase 的高可用就是通过 ZooKeeper 来实现的。

（4）数据迁移

在数据应用中，通常会有不同系统间的数据迁移操作。在 Hadoop 生态圈中，Sqoop 和 Flume 框架可以很好地解决不同系统间的数据收集和传输。

Sqoop 是一款开源的工具，主要用在关系型数据库、数据仓库和 Hadoop 之间进行数据迁移。在实际应用中，可以使用 Sqoop 完成关系型数据库到 HDFS、Hive 等框架的数据导入导出操作。

Flume 是 Cloudera 提供的一个高可用、高可靠、分布式的框架，主要用于分布式海量日志数据的高效搜集、聚合和传输。Flume 支持在日志系统中定制各类数据发送方，用于收集数据，同时 Flume 提供对数据进行简单处理并写入各种数据接收方的能力。

（5）NoSQL

在 Hadoop 生态圈中，HBase 提供了 NoSQL 数据库的功能，用于满足大数据应用中快速随机访问大数据量（PB 量级）数据并及时响应用户的需求。

HBase 是建立在 HDFS 上的面向列的 NoSQL 数据库，可以对大规模数据进行随机、实时读/写访问。HBase 具有可伸缩、高可靠、高性能的特点。

（6）机器学习

目前，机器学习已经发展成为一个热门话题，Hadoop 生态圈中提供了 Mahout 库来完成机器学习功能。

Mahout 提供了一些可扩展的机器学习领域经典算法的实现，目的是帮助开发人员方便快捷地开发智能应用程序。Mahout 是一个机器学习和数据挖掘库，包括聚类、分类、推荐引擎（协同过滤）等数据挖掘方法，用户可以通过调用算法包来缩短编程时间。

（7）任务调度

大数据实际应用中，通常会遇到多个作业协同完成一个业务分析的场景。这就需要一个能合理管理作业调度的框架，在 Hadoop 生态圈中，由 Oozie 负责解决任务调度问题。

Oozie 是一个工作流引擎，是基于 Hadoop 的调度器，可以调度 MapReduce、Pig、Hive、Spark 等不同类型的单一或者具有依赖性的作业。当一个作业中需要调用几个任务时，就可以使用 Oozie 将任务串联，再通过 Oozie 调度整个任务进程。

1.2.3 Hadoop 应用案例

1．大数据在传媒行业的应用

大数据正逐渐上升为不同新闻媒体晋升一流媒体的优化路径，如何利用缜密的大数据思维和良好的大数据洞察力推动传媒生态升级转型，使自身完全具备大数据应用的能力，已被各大媒体提上重要议程。大数据时代正在推动整个传媒行业发生深刻变革，要想在这场变革中获得优势，必须构建创新思维体系，充分利用传统媒体和大数据时代的特点，发展壮大传媒行业。

通过全网舆情数据汇聚互联网上的文章内容，挖掘全网最热新闻话题和话题评价，为新闻发稿、栏目制作做参考，达到报道分析热点的抓取、节目制作的改进、明星邀约的助攻。通过智能算法得出有可能成为明日头条的新闻事件，实现最快新闻发稿，最贴合节目编辑，达到快人一步、领先全部。

通过创建微博分析任务进行传播分析，展现微博曝光数，挖掘话题的兴趣图谱、普通用户与大 V 的关系网络，整合标签进行多维透视，深度刻画人群画像，精准触达目标客户。

2．大数据在智能交通领域的应用

近年来，随着我国经济的快速发展，机动车持有量迅速增加，交通管理现状和需求的矛盾进一步加剧。在此情况下，如何利用先进的科技手段提高交通管理水平，抑制交通事故发生，是当前交通管理部门亟待解决的问题。

针对交通管理部门的需求以及我国的道路特点，可通过整合图像处理、模式识别等技术，对监控路段的机动车道、非机动车道进行全天候实时监控和数据采集。前端卡口处理系统对拍摄的图像进行分析，获取车牌号码、车牌颜色、车身颜色、车标、车辆子品牌等数据，并将获取到的车辆信息连同车辆的通过时间、地点、行驶方向等信息，通过计算机网络传输到卡口系统控制中心的数据库中，进行数据存储、查询、比对等处理，当发现肇事逃逸、违章或可疑车辆时，系统会自动向拦截系统及相关人员发出告警信号，为交通违章处理、交通事故逃逸、盗抢机动车辆等案件的及时侦破提供重要的信息和证据。同时，随着全城 Smart 系统的建设，新型的 Smart IPC 监控前端也将成为一个卡口系统，这使得城市卡口系统更加严密，能够获取到更多的过往车辆数据，能够更准确地描绘出车辆动态信息。

基于大数据的智慧交通存在多种可能。交通的智能化是根本的趋势，利用大数据技术和智能分析技术，整合城市管理的其他数据，将真正推动智慧交通建设，为交通管理奠定良好的基础。目前大数据技术主要应用在交管部门所辖道路，随着数据的进一步联网开放，可以整合停车场、铁路、轨道交通、公交等各种来源的数据，提供更为丰富的城市交通应用，让道路更加畅通，停车位不再难找，提升城市交通整体运营效率。

3. **大数据在金融行业的应用**

下面从四个方面介绍大数据技术在金融行业的应用。

（1）客户画像应用

在银行业务中，银行拥有的客户信息并不全面，基于银行自身拥有的数据有时难以得出理想的结果甚至可能得出错误的结论。比如，某位信用卡客户月均刷卡 8 次，平均每次刷卡金额 800 元，平均每年拨打 4 次客服电话，从未有过投诉。按照传统的数据分析，该客户应该是一位满意度较高、流失风险较低的客户。但如果看到该客户的微博，看到的真实情况却是：由于工资卡和信用卡不在同一家银行，导致还款不方便，该客户好几次打客服电话没接通，并多次在微博上抱怨。可见该客户的流失风险较高。所以银行不仅要考虑银行自身业务采集到的数据，还应考虑整合更多的外部数据，可以使用大数据客户画像方式采集客户特征、客户标签，完成有效的客户画像，便能对客户进行有效的数据建模和有效准确的营销。

（2）精准营销

在客户画像的基础上银行可以有效地开展精准营销，包括：

实时营销：根据客户的实时状态来进行营销。

交叉营销：即不同业务或产品的交叉推荐。

个性化推荐：根据客户的喜好进行服务或者进行产品的个性化推荐。

客户生命周期的管理：新客户获取、老客户流失和老客户赢回等。

（3）风险管控

风险管控包括中小企业贷款风险评估、欺诈交易识别和反洗钱等手段。

中小企业贷款风险评估：将企业的产品、流通、销售、财务等信息结合大数据挖掘方法进行贷款风险分析。

欺诈交易识别和反洗钱：利用持卡人基本信息、交易模式、行为模式等，结合智能规则进行交易反欺诈分析。

（4）运营优化

市场和渠道分析优化：通过大数据，银行可以监控不同市场推广渠道尤其是网络渠道推广的质量，从而进行合作渠道的调整和优化。

产品和服务优化：金融行业可以将客户行为转化为信息流，并从中分析客户的个性特征和风险偏好，更深层次地理解客户习惯，智能化分析和预测客户需求，从而进行产品创新和服务优化。

舆情分析：金融行业可以通过爬虫技术，抓取社区、论坛和微博上关于银行以及银行产品和服务的相关信息，并通过自然语言处理技术进行正负面舆情判断。

任务 3　搭建移动通信业务的 Hadoop 处理平台

【任务描述】

搭建 Hadoop 伪分布式环境。

【关键步骤】

（1）安装虚拟机。

（2）安装 Linux 操作系统。

（3）搭建 Hadoop 伪分布式环境。

1.3.1　安装虚拟机

1. 虚拟机概述

虚拟机是指通过软件模拟的具有完整硬件系统功能的、运行在一个完全隔离环境中的完整计算机系统。虚拟机软件允许用户在一台机器上同时运行多个不同类型的操作系统，可以模拟一个标准的计算机环境，包括 CPU、内存、显卡、硬盘、网卡、声卡、USB 控制器等。

目前，流行的虚拟机软件有 VMware Workstation 和 VirtualBox，它们都能在 Windows 系统上虚拟出多个计算机。由于 VMware 的功能更加完善，所以本书采用 VMware Workstation 虚拟机，版本选择 VMware Workstation 12。读者可以在官网下载对应的版本。

2. VMware 安装

安装 VMware Workstation 的过程如下。

（1）双击下载的安装文件，进入到安装向导界面，如图 1.2 所示。

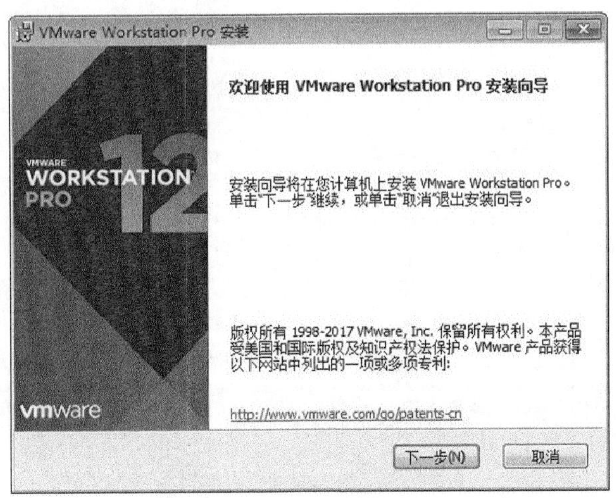

图1.2　VMware安装向导界面

（2）在安装向导界面，单击“下一步”按钮，进入到最终用户许可协议界面，如图 1.3 所示。

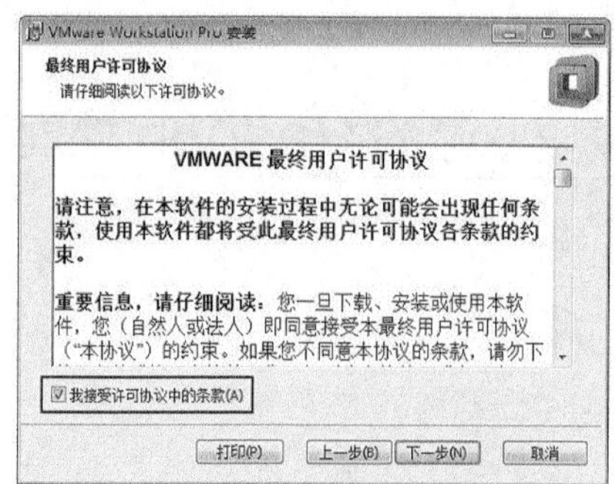

图1.3　VMware许可协议界面

（3）选中“我接受许可协议中的条款”复选框，单击“下一步”按钮，选择“自定义”单选按钮进入自定义安装界面，如图 1.4 所示。

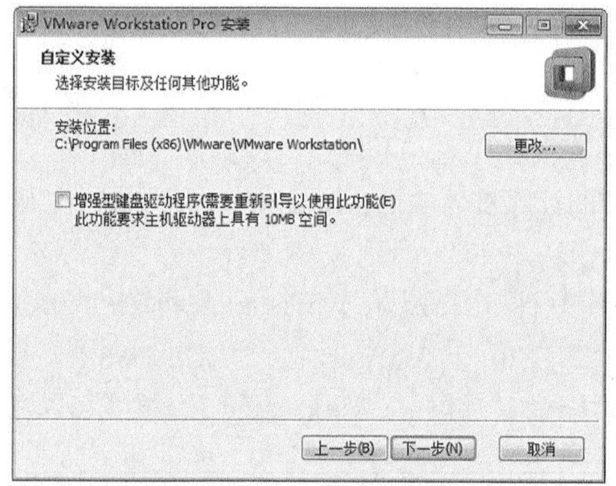

图1.4　VMware自定义安装界面

（4）单击“下一步”按钮进入快捷方式选择界面，勾选“桌面”和“开始菜单程序文件夹”复选框后单击“下一步”按钮，如图 1.5 所示。

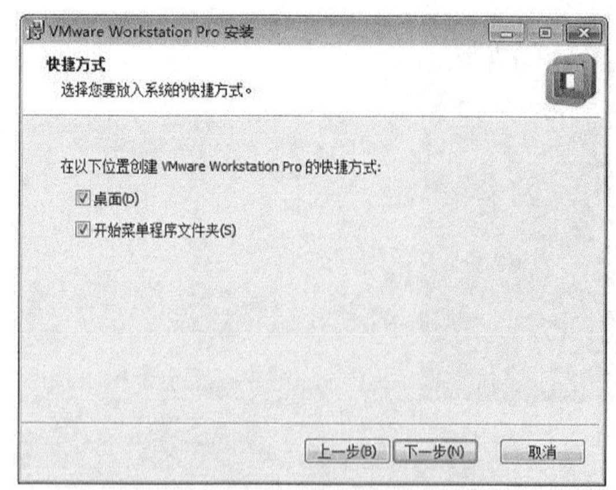

图1.5　VMware快捷方式选择界面

（5）单击“安装”按钮即可完成安装，如图 1.6 所示。

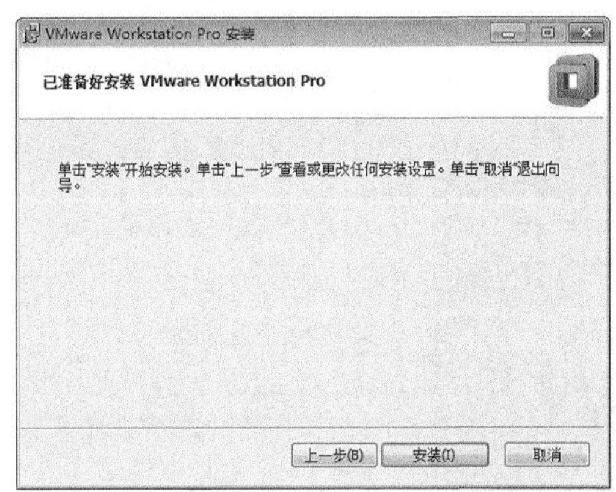

图1.6 VMware准备安装界面

1.3.2 安装 Linux 操作系统

1. Linux 概述

Linux 是一套免费使用和自由传播的类 UNIX 操作系统，是一个基于 POSIX 和 UNIX 的多用户、多任务、支持多线程和多 CPU 的操作系统。Linux 操作系统诞生于 1991 年 10 月 5 日，可安装在手机、平板电脑、台式计算机、大型服务器等设备中。目前，大多数企业选择 Linux 作为服务器的操作系统。

Linux 存在很多变种以及版本。

（1）Ubuntu。2004 年 9 月发布，是最为流行的桌面 Linux 发行版本，个人用户使用较多，社区很庞大。

（2）Red Hat。使用最广泛，性能比较稳定，属于商业版本。

（3）CentOS。2003 年年底发布，是对商业版 RHEL（Red Hat Enterprise Linux）的重新编译，免费开源，性能稳定。目前主流企业仍旧选用 Red Hat 或者 CentOS，本书选择 CentOS 7 版本的 iso 镜像文件（CentOS-7-x86_64-DVD-1804.iso）。读者可以在 CentOS 官网下载对应的版本。

2. CentOS 安装

（1）打开安装完成的 VMware 虚拟机，单击【文件】/【新建虚拟机】或直接单击【创建新的虚拟机】图标，如图 1.7 所示。

图1.7 新建虚拟机

（2）选择“自定义”单选按钮，单击“下一步”按钮，如图 1.8 所示。

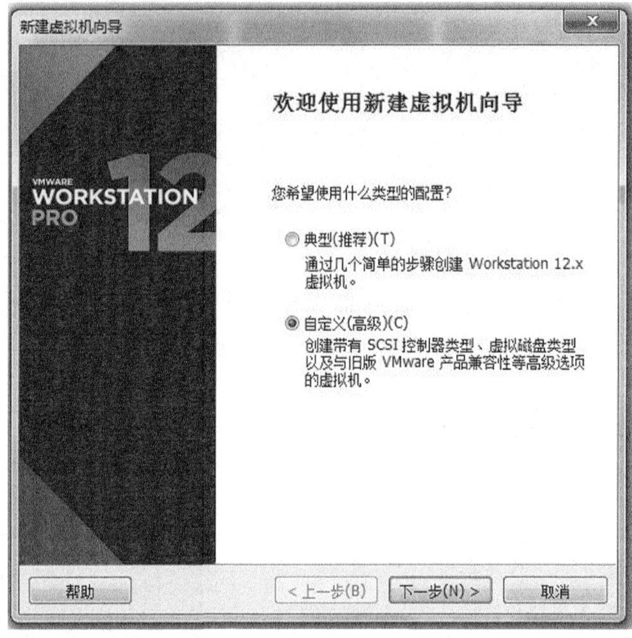

图1.8　自定义安装

（3）选择虚拟机硬件兼容性，如图 1.9 所示。

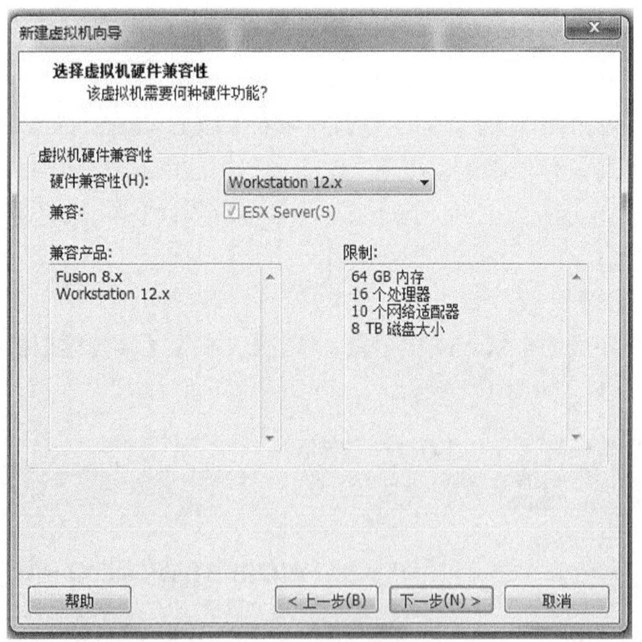

图1.9　选择硬件兼容性

（4）选择“稍后安装操作系统”，如图 1.10 所示。

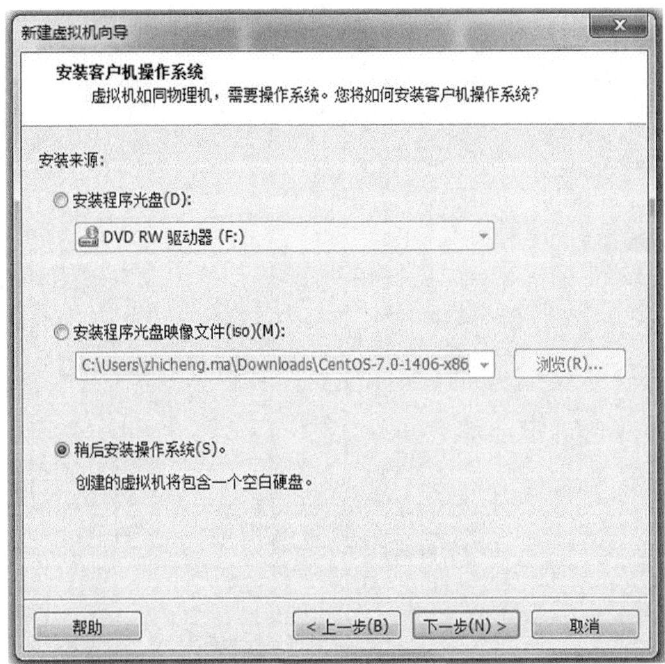

图1.10　选择安装来源

（5）选择客户机操作系统和版本，如图 1.11 所示。

图1.11　选择Linux以及CentOS 64位操作系统

（6）输入虚拟机名称和安装路径，如图 1.12 所示。

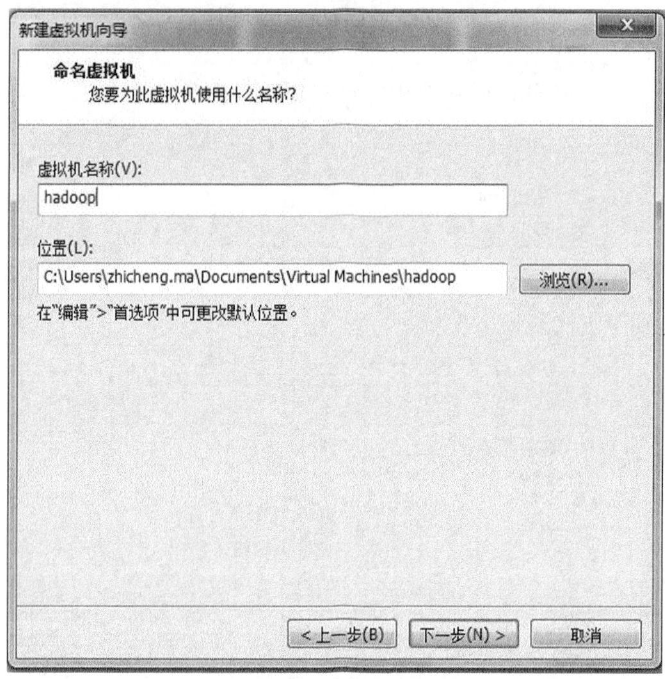

图1.12 选择虚拟机名称和安装路径

（7）选择“使用网络地址转换（NAT）”，如图 1.13 所示。

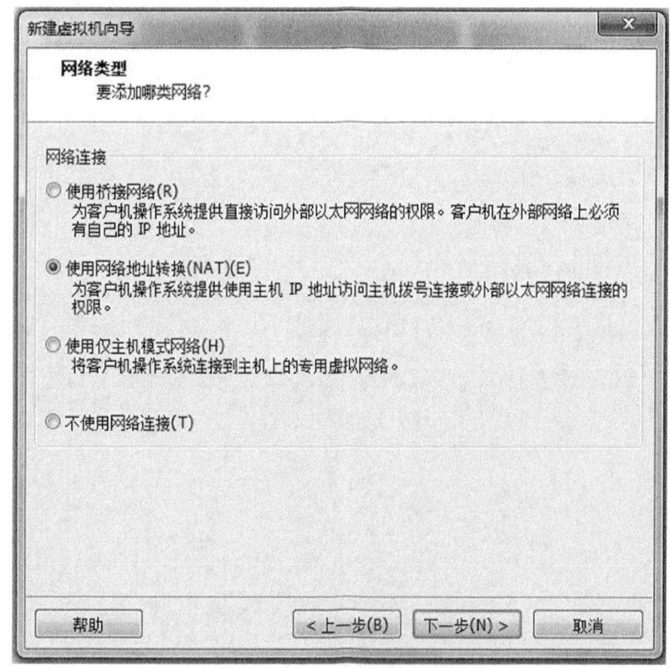

图1.13 选择网络类型

（8）选择“创建新虚拟磁盘”，如图 1.14 所示。

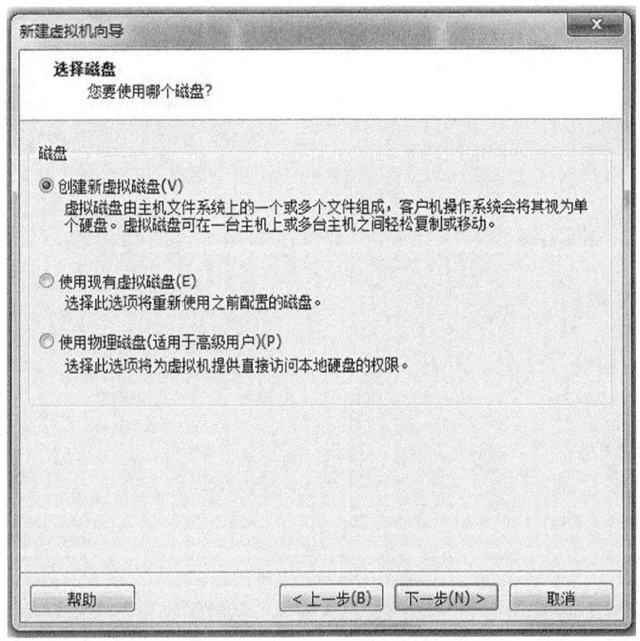

图1.14　选择磁盘

（9）指定磁盘容量。分别选择最大磁盘大小为 20GB 和“将虚拟磁盘拆分成多个文件”单选按钮，如图 1.15 所示。

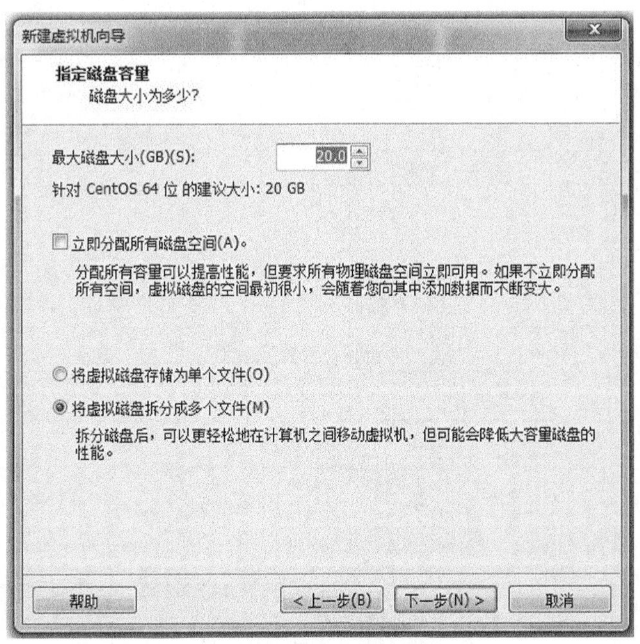

图1.15　指定磁盘容量

（10）磁盘文件选择默认即可，如图 1.16 所示。

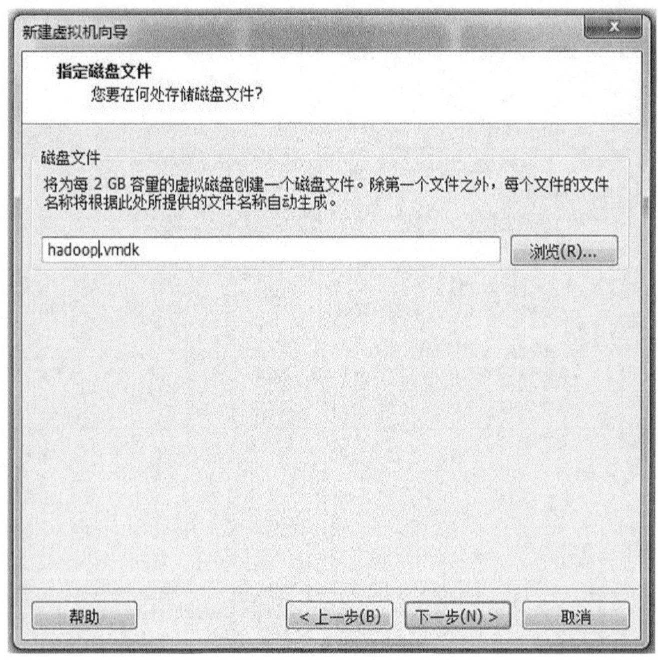

图1.16　指定磁盘文件

（11）单击“自定义硬件”按钮，如图 1.17 所示。

图1.17　自定义硬件

（12）选择“新 CD/DVD”和“使用 ISO 映像文件”单选按钮，浏览找到本地镜像下载位置，如图 1.18 所示。

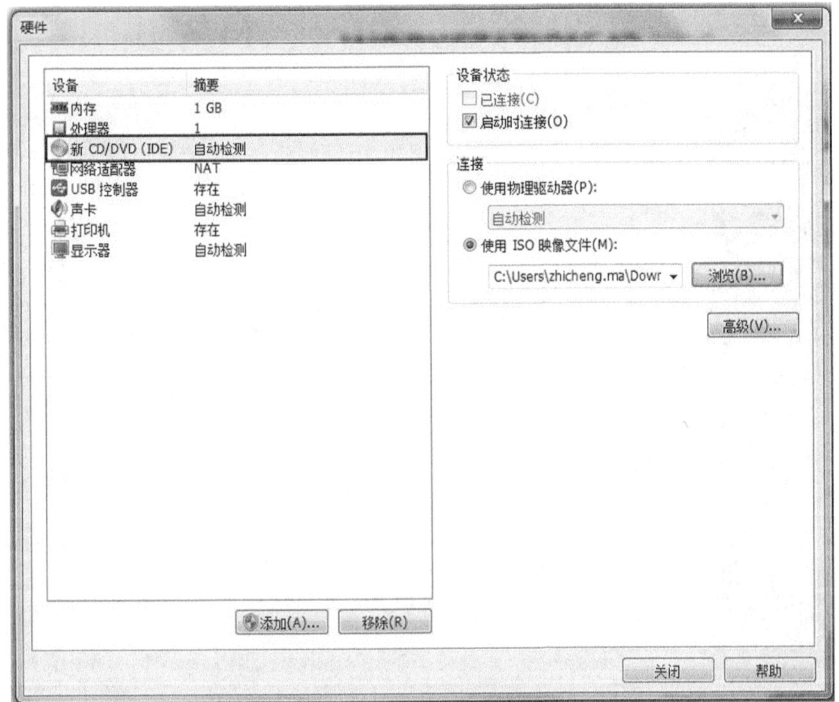

图1.18　选择镜像位置

（13）单击“完成”按钮即可。

3. 启动虚拟机安装操作系统

（1）进入 CentOS 7 安装欢迎界面，如图 1.19 所示。

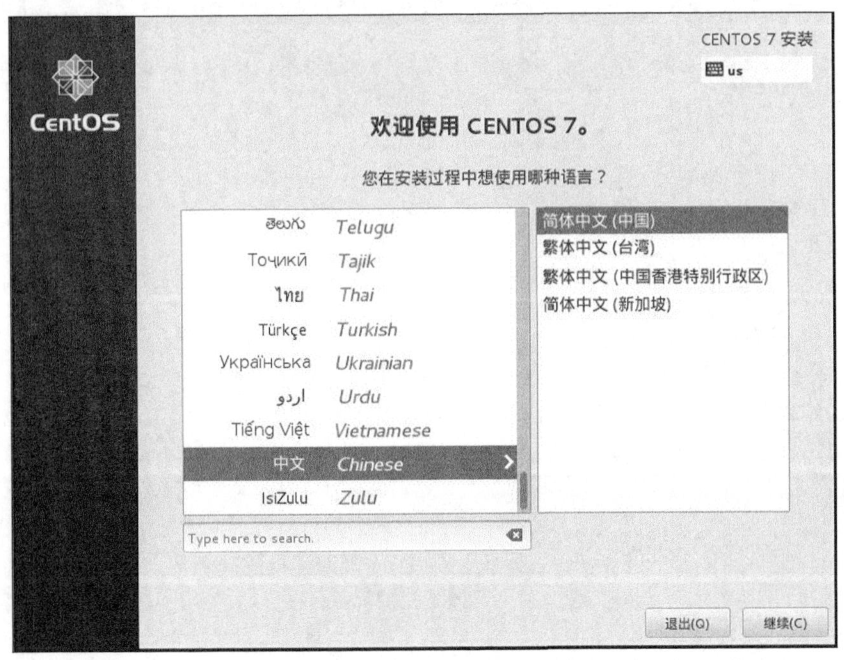

图1.19　CentOS 7安装欢迎界面

（2）选择安装位置，如图 1.20 所示。

图1.20　选择安装位置

（3）选择“我要配置分区”，如图 1.21 所示。

图1.21　选择“我要配置分区”

（4）在“手动分区”界面选择“标准分区”，如图 1.22 所示。

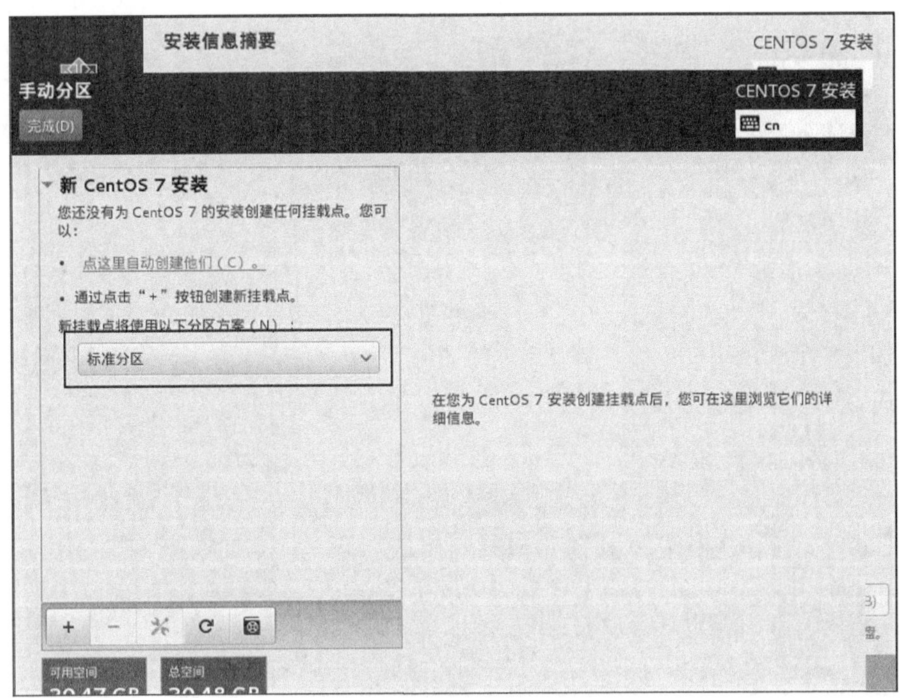

图1.22　选择"标准分区"

（5）选择 swap 分区，"期望容量"大小为 4096，如图 1.23 所示。

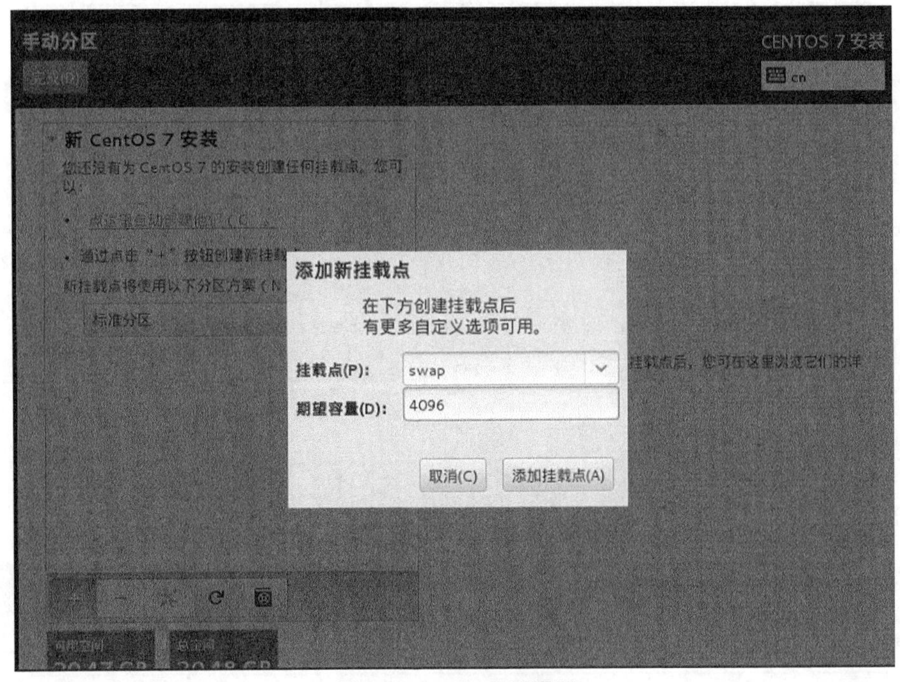

图1.23　swap分区配置

（6）选择根分区，期望容量为空，如图 1.24 所示。

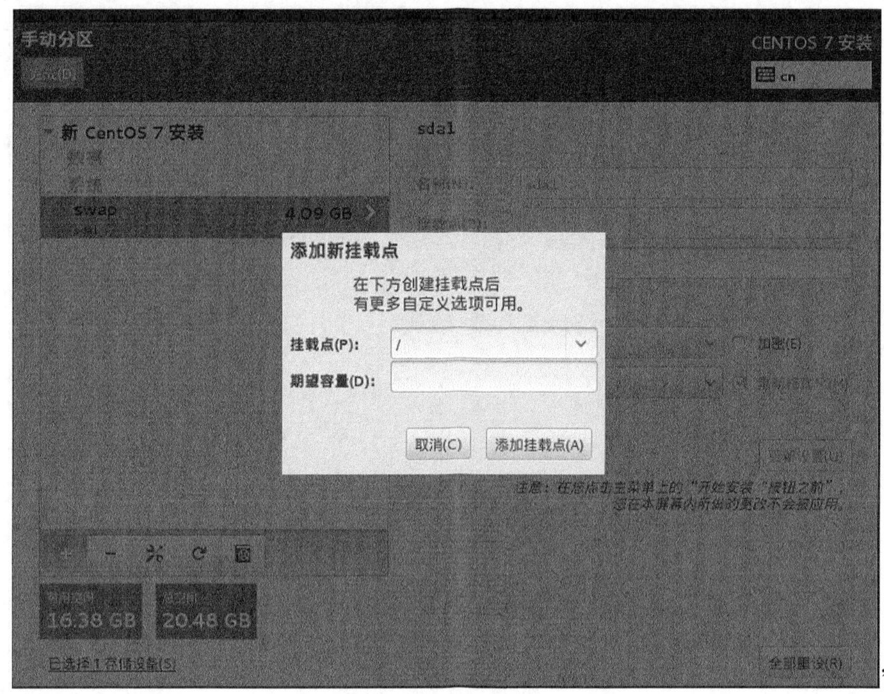

图1.24　根分区配置

（7）在设置完成界面单击左上角的“完成”按钮即可，如图 1.25 所示。

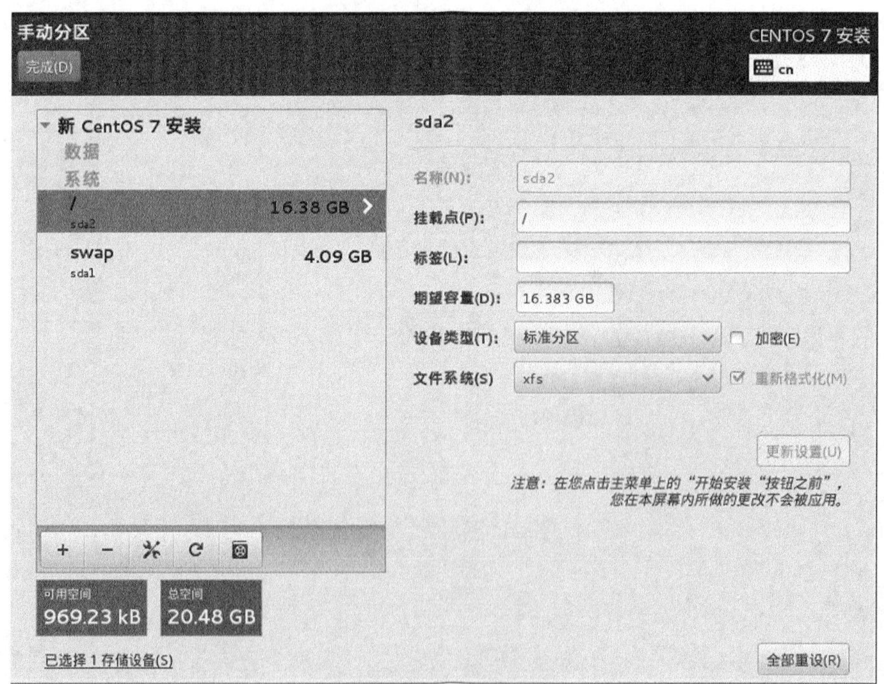

图1.25　手动分区

（8）选择“接受更改”按钮，如图 1.26 所示。

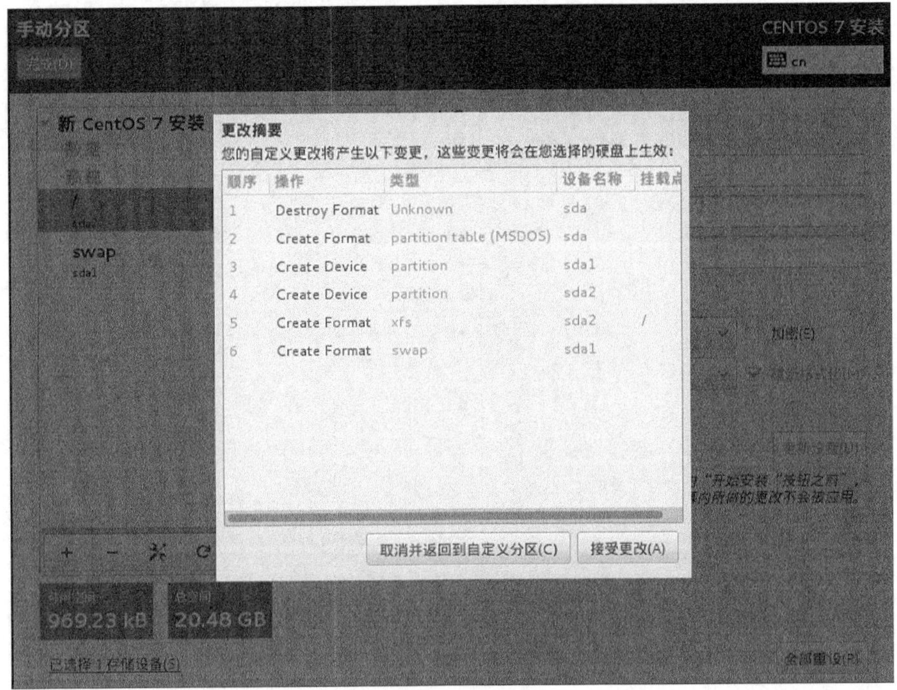

图1.26　接受更改

（9）返回到 CentOS7 安装界面，设置“网络和主机名”，如图 1.27 所示。

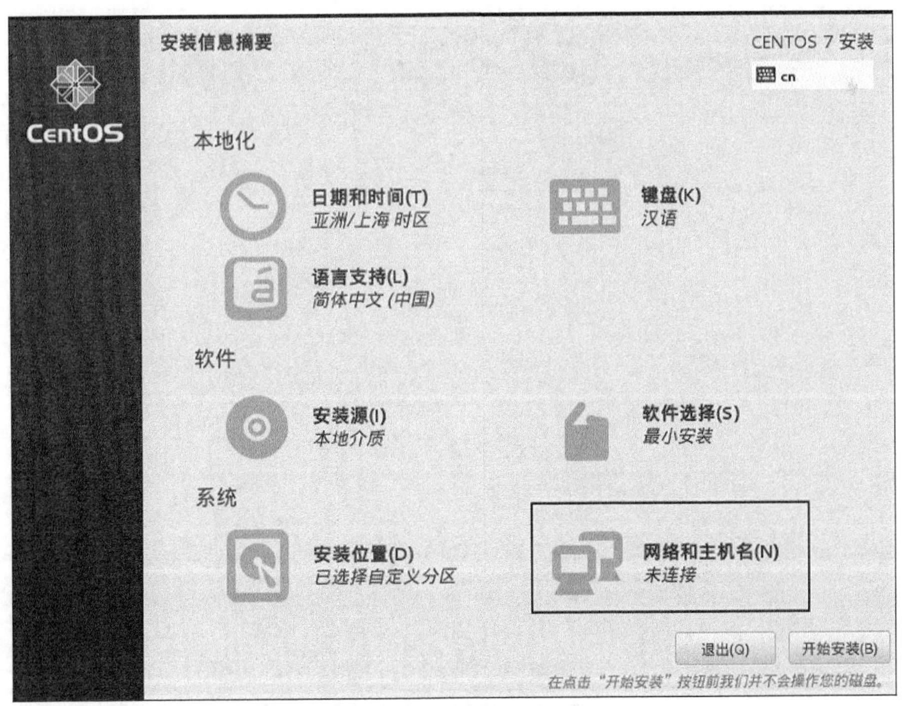

图1.27　设置网络和主机名

（10）开启网络配置，修改主机名为 hadoop，如图 1.28 所示。

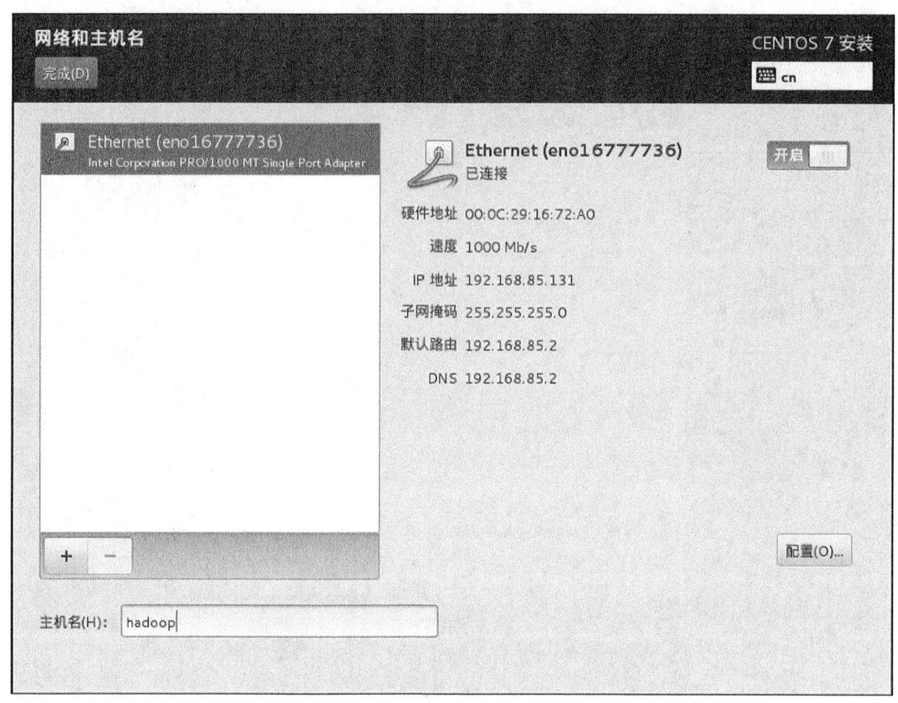

图1.28　开启网络配置和修改主机名

（11）为 root 账户设置密码，如图 1.29 所示。

图1.29　设置root账户密码

（12）创建一个新用户 hadoop 并设置密码，如图 1.30 所示。

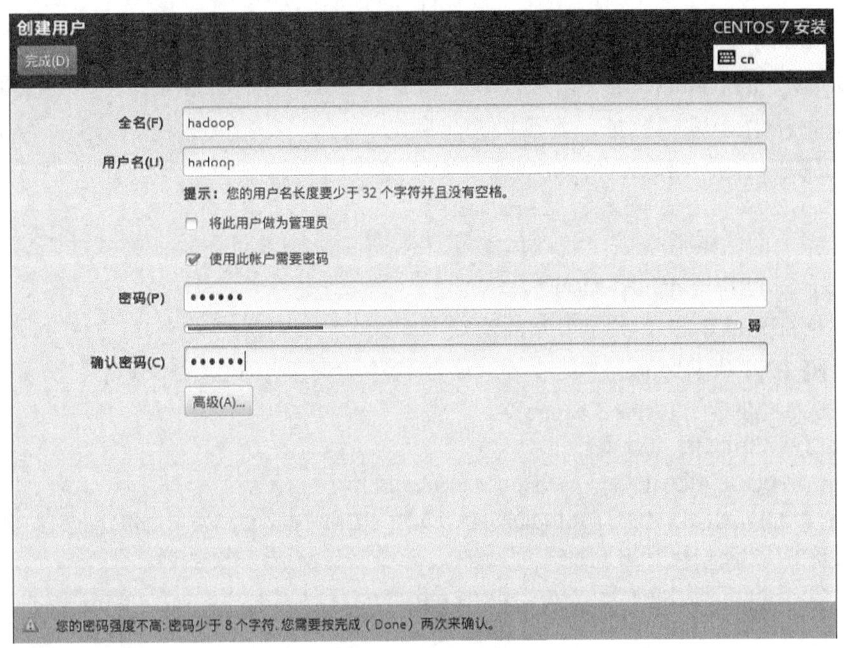

图1.30　创建新用户

（13）重启虚拟机即可完成系统新建。

Hadoop
伪分布式
环境搭建

1.3.3　搭建移动通信业务的 Hadoop 处理平台

安装完虚拟机和 CentOS 操作系统以后，相当于准备好了服务器环境，接下来就要在服务器环境上安装处理移动通信业务的 Hadoop 处理平台。安装 Hadoop 伪分布式环境的详细步骤请扫描二维码。

1. **安装包下载**

本书采用的是 CDH 版本的 Hadoop-2.6.0-cdh5.14.2，读者可以到 Cloudera 官网去下载对应的版本。由于 Hadoop 是依赖于 JDK 的，所以也需要下载 JDK。将软件下载到合适的位置，本书选择下载到安装用户“hadoop”home 目录的 software 子目录下。

2. **解压安装**

下载完成以后，需要将安装包解压。

```
[hadoop@hadoop ~]$ tar –zxvf /home/hadoop/software/hadoop-2.6.0-cdh5.14.2.tar.gz
```

将解压后的文件复制到/opt 目录下。

```
[hadoop@hadoop ~]$ sudo mv hadoop-2.6.0-cdh5.14.2 /opt/hadoop-2.6.0-cdh5.14.2
```

注意

由于使用的是普通用户进行安装，所以需要加 sudo 才能操作/opt 目录。

JDK 的安装可参考 Hadoop 的解压安装方式自己完成。

3. Hadoop 伪分布式环境搭建

Hadoop 中的 NameNode 和 DataNode 是通过 SSH（Secure Shell）进行通信的，所以需要先完成 SSH 免密码登录。具体操作步骤如下。

```
[hadoop@hadoop ～]$ ssh-keygen -t rsa
[hadoop@hadoop ～]$ ssh-copy-id localhost
```

执行完上面两条命令后，就可以实现免密码登录到本机。验证方式如下。

```
[hadoop@hadoop ～]$ ssh hadoop
```

前置条件完成以后，就开始准备搭建 Hadoop 平台，步骤如下。

（1）将 JDK 和 Hadoop 的安装目录添加到环境变量当中，命令如下。

```
[hadoop@hadoop ～]$ vi ～/.bashrc
export JAVA_HOME=/opt/jdk1.8.0_171
export HADOOP_HOME=/opt/hadoop-2.6.0-cdh5.14.2
export PATH=.:$HADOOP_HOME/bin:$HADOOP_HOME/sbin:$JAVA_HOME/bin:$PATH
```

修改完成以后，执行 source ～/.bashrc 命令，让环境变量生效。

（2）在 Hadoop 环境配置文件$HADOOP_HOME/etc/hadoop/hadoop-env.sh 中添加以下代码。

```
export JAVA_HOME= /opt/jdk1.8.0_171
```

（3）在 Hadoop 的核心配置文件 core-site.xml 中添加以下代码。

```
//配置 NameNode 的主机名和端口号
<property>
        <name>fs.defaultFS</name>
        <value>hdfs://hadoop:8020</value>
</property>
```

（4）在 HDFS 的核心配置文件 hdfs-site.xml 中添加以下代码。

```
//设置 HDFS 元数据文件存放路径
<property>
        <name>dfs.namenode.name.dir</name>
        <value>/home/hadoop/tmp/dfs/name</value>
</property>
//设置 HDFS 数据文件存放路径
<property>
        <name>dfs.datanode.data.dir</name>
        <value>/home/hadoop/tmp/dfs/data</value>
</property>
//设置 HDFS 文件副本数
<property>
        <name>dfs.replication</name>
        <value>1</value>
</property>
//设置其他用户执行操作时会提醒没有权限
<property>
        <name>dfs.permissions</name>
```

```
        <value>false</value>
</property>
```

（5）在 MapReduce 配置文件 map-site.xml 中添加如下代码。

```
<property>
        <name>mapreduce.framework.name</name>
        <value>yarn</value>
</property>
```

（6）在 YARN 配置文件 yarn-site.xml 中添加如下代码。

```
<property>
        <name>yarn.nodemanager.aux-services</name>
        <value>mapreduce_shuffle</value>
</property>
```

（7）在从节点配置文件 slaves 中添加以下代码。

```
hadoop
```

4. 格式化 HDFS 系统

Hadoop 配置完成以后，第一次使用 Hadoop 平台需要先格式化文件系统。

示例 1

格式化移动通信业务处理平台的 HDFS 系统。

```
[hadoop@hadoop ~]$ hdfs namenode -format
```

5. 启动 Hadoop 平台

Hadoop 的启动，通常有两种方式。

（1）一次启动所有进程

示例 2

启动移动通信业务处理平台的全部进程。

```
[hadoop@hadoop ~]$ $HADOOP_HOME/sbin/start-all.sh
```

（2）单独启动每个进程

示例 3

在移动通信业务处理平台上启动 HDFS。

```
[hadoop@hadoop ~]$ $HADOOP_HOME/sbin/start-dfs.sh
```

示例 4

在移动通信业务处理平台上启动 YARN。

```
[hadoop@hadoop ~]$ $HADOOP_HOME/sbin/start-yarn.sh
```

注意

单独启动进程的时候，需要先启动 HDFS，再启动 YARN。

6. 验证

Hadoop 启动成功以后，通常使用以下几种方式进行验证。

（1）通过 jps 命令查看启动的进程。命令如下。

```
[hadoop@hadoop ~]$ jps
```

命令执行完成以后，可以看到 NameNode、DataNode、SecondaryNameNode、ResourceManager、NodeManager 进程，就证明 Hadoop 平台已经启动成功。

（2）通过 webui 的方式验证。

验证 HDFS 启动成功的网址为：http://hadoop:50070。

验证 YARN 启动成功的网址为：http://hadoop:8088。

两个网址如果都能正确打开，也证明 Hadoop 平台已经启动成功。

（3）运行 Hadoop 平台自带的 wordcount 示例。

Hadoop 安装包自带了很多应用程序，包括 wordcount（统计文件中单词出现的次数）、PI（计算圆周率 π 的值）等应用程序，用户可以直接运行这些应用程序。存放的路径为：$HADOOP_HOME/share/hadoop/mapreduce/hadoop-mapreduce-examples-2.6.0-cdh5.14.2.jar。

示例 5

在移动通信业务处理平台上运行 wordcount 程序。

① 先在 HDFS 上创建几个数据目录。

```
[hadoop@hadoop ~]$ hdfs dfs -mkdir -p /data/wordcount
[hadoop@hadoop ~]$ hdfs dfs -mkdir -p /output/
```

② 创建 inputWord 文件并输入测试数据，使用制表符对数据进行分隔。

```
[hadoop@hadoop ~]$ vi ~/inputWord
```

测试数据如下：

```
hello hadoop	hello
hello welcome	hadoop
```

③ 上传文件。

```
[hadoop@hadoop ~]$ hdfs dfs -put inputWord /data/wordcount/
```

④ 提交作业。

```
[hadoop@hadoop ~]$ hadoop jar /home/hadoop/app/hadoop-2.6.0-cdh5.14.2/share/hadoop/mapreduce/
hadoop-mapreduce-examples-2.6.0-cdh5.14.2.jar wordcount /data/wordcount /output/wordcount
```

⑤ 查看结果。

```
[hadoop@hadoop ~]$ hdfs dfs -text /output/wordcount/part-r-00000
hello	3
welcome	2
hadoop	2
```

1.3.4 大数据集群管理平台

1. 简介

为了能让用户更好地使用 Hadoop 生态圈技术，更方便地管理 Hadoop 集群，很多公司都推出了集群管理平台，最具代表性的就是 Cloudera 公司的 Cloudera Manager 管理平台和 Apache Ambari 管理平台。这些平台不仅节约了部署时间，还提供了一系列监控和诊断工具，可以帮助用户优化集群性能，提供实时的集群概况。由于本书采用的软件都是 CDH 的版本，所以只针对 Cloudera Manager 进行介绍。

Cloudera Manager 是 Cloudera 公司推出的管理平台，对 CDH 的每个部件都提供了细粒度的可视化和控制。Cloudera Manager 设计的目的是使企业数据中心的管理变得简单和直观。通过 Cloudera Manager，可以方便地部署并且集中式地操作完整的大数据软件栈，还可以自动化安装过程，从而减少部署集群的时间。

2. Cloudera Manager 下载安装

Cloudera Manager 可以通过官网进行下载，下载完成后，就可以安装 Cloudera Manager 了。Cloudera Manager 的具体安装步骤可扫描二维码了解。

Cloudera Manager 集群管理平台搭建

1.3.5 技能实训

读者可按照任务 3 中介绍的步骤，搭建自己的 Hadoop 平台。

实现步骤：

（1）安装虚拟机和 CentOS。

（2）安装 JDK 和 Hadoop 软件。

（3）完成 SSH 免密码登录操作。

（4）配置 Hadoop 平台相关配置文件。

（5）格式化文件系统。

（6）启动 Hadoop 平台。

本章总结

➢ 在大数据领域，可以使用 Hadoop 及 Hadoop 生态圈技术来解决大数据存储和计算问题。

➢ 掌握搭建 Hadoop 伪分布式环境的步骤。

本章作业

一、简答题

1. 简述 hdfs-site.xml 配置文件中需要配置的三个属性名称及其含义。

2. 写出 Hadoop 的三种运行模式。

二、编码题

1. 在 Hadoop 平台上运行自带的测试案例 PI，计算 π 的值。

提示：可以使用如下命令查看 PI 的使用方法。

hadoop jar $HADOOP_HOME/share/hadoop/mapreduce/hadoop-mapreduce-examples-2.6.0-cdh5.14.2.jar PI。

2. 在 Hadoop 平台上运行自带的测试案例 wordmean，用于计算文件中单词的平均长度。

提示：可以使用如下命令查看 wordmean 的使用方法。

hadoop jar $HADOOP_HOME/share/hadoop/mapreduce/hadoop-mapreduce-examples-2.6.0-cdh5.14.2.jar wordmean。

3．在 Hadoop 平台上运行自带的测试案例 wordmedian，用于计算文件中单词的中位长度。

提示：可以使用如下命令查看 wordmedian 的使用方法。

hadoop jar $HADOOP_HOME/share/hadoop/mapreduce/hadoop-mapreduce-examples-2.6.0-cdh5.14.2.jar wordmedian。

第 2 章

Hadoop 分布式文件系统 HDFS

技能目标

- 理解 HDFS 的体系架构
- 掌握 HDFS 的访问方式
- 掌握 HDFS 文件的读写流程
- 了解 Hadoop 序列化机制

本章任务

任务 1　了解 HDFS
任务 2　使用 HDFS 处理移动通信数据文件
任务 3　了解 HDFS 运行原理
任务 4　实现移动通信数据的行文件方式存储

本章资源下载

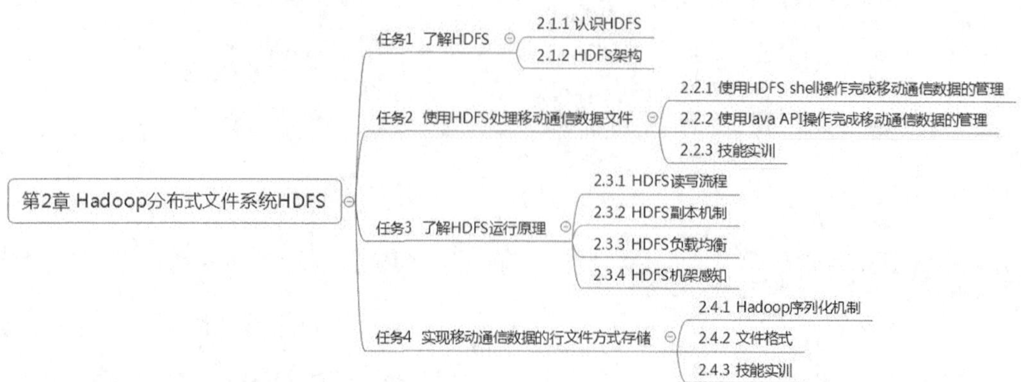

Hadoop 分布式文件系统（Hadoop Distributed File System，HDFS）是 Apache 顶级项目 Hadoop 的一个核心构成，由于 HDFS 可以部署在普通硬件设备上，因此大多数企业都选择 HDFS 作为大数据业务的存储系统。本章主要讲解 HDFS 的体系架构、通过 shell 操作命令管理数据、HDFS 的运行原理、序列化机制、常用文件格式等。

任务 1 了解 HDFS

【任务描述】

了解 HDFS 的产生背景、HDFS 的特点和设计目标，掌握 HDFS 架构的核心组成。

【关键步骤】

（1）认识 HDFS。

（2）了解 HDFS 的优缺点。

（3）掌握 HDFS 的体系架构。

2.1.1 认识 HDFS

1. HDFS 产生背景

当今世界正处在大数据的时代，数据不仅由互联网产生，科学计算、物流、工业设备、道路交通等领域也在产生海量的数据。随着数据量越来越大，使用单个操作系统的存储方式显然已经不能满足大数据存储的需求，因此，迫切需要一种系统来存储大数据时代下产生的海量数据，于是分布式文件系统（Distributed File System，DFS）就诞生了。

分布式文件系统是指文件系统管理的物理资源不一定直接连接在本地节点上，而是通过计算机网络与节点相连。它允许将一个文件通过网络在多台主机上以多副本（提高容错性）的方式进行存储，实际上是通过网络来访问文件，而用户和程序看起来却像是访问本地的文件系统一样。

2. HDFS 简介

HDFS 是 Hadoop 项目的核心子项目，用于大数据领域的数据存储。HDFS 是被设计成适合运行在通用硬件（commodity hardware）上的分布式文件系统。它与现有的分布式文件系统有很多近似的地方，也有很多明显的不同。HDFS 具有高容错、高可靠性、高可扩展性、高获得性、高吞吐量等特征。

3. HDFS 优缺点

（1）HDFS 的优点

➢ 支持处理超大文件。这里的超大文件通常是指 GB 到 TB 量级的数据文件，Hadoop 并不怕文件大；相反，如果 HDFS 中存在众多的小文件，反而会导致 Hadoop 集群的性能有所下降。

➢ 运行在廉价的机器上。Hadoop 集群可以部署在普通的廉价机器之上，而无需部署在价格昂贵的小型机器上，这样可以降低公司的运营成本。

➢ 高容错性。HDFS 上传的数据会自动保存多个副本，即通过增加副本的数量来增加 HDFS 的容错性。一个副本丢失，HDFS 副本冗余机制会自动复制其他机器上的副本。

➢ 流式文件写入。HDFS 提供一次写入、多次读取的服务。文件一旦写入，就不能修改，只能增加，可以提高 I/O 性能。

（2）HDFS 的缺点

➢ 不适合低延迟数据访问场景。HDFS 本身是为存储大数据而设计的，对延时要求在毫秒级别的应用，不适合采用 HDFS。实时性、低延迟的查询应用 HBase 会是更好的选择。

➢ 不适合小文件存储场景。HDFS 中的元数据（如目录结构、文件目录属性、文件 Block 的节点列表等）存放在 NameNode 中，整个文件系统的文件数量会受限于 NameNode 的内存大小。一旦集群中的小文件过多，会导致 NameNode 的压力倍增，进而影响到集群的性能。一般采用 SequenceFile 等方式对小文件进行合并，或者是使用 NameNode Federation 的方式来改善。

➢ 不适合并发写入，文件随机修改场景。HDFS 采用追加（append-only）的方式写入数据，不支持在文件的任意位置修改，写操作总是在文件的末尾进行。

4. HDFS 设计目标

有关 HDFS 设计目标的详细描述可以参考 Hadoop 的官网文档。本节只挑选三个重要的设计目标进行讲解。

（1）硬件故障。硬件故障是常态而不是异常。整个 HDFS 系统由成百上千个服务器所构成，每个服务器上存储着文件系统的部分数据。构成系统的组件数目是非常巨大的，任一个组件都有可能失效，这意味着总有一部分 HDFS 组件是不工作的。因此故障检测和自动快速恢复是 HDFS 最核心的架构设计目标。

（2）大规模数据集。运行在 HDFS 上的应用具有很大的数据集，这意味着典型的 HDFS 文件大小一般都在 GB 至 TB 量级。HDFS 应该提供很高的聚合数据带宽，能在一个集群里扩展到数百个节点。一个单一的 HDFS 实例应该能支撑数以千万计的文件。

（3）移动计算比移动数据更经济。在靠近计算的数据存储的位置进行计算是最理想的状态，尤其是在数据达到海量数据级别时更是如此。这样不仅消除了网络的拥堵，而且提高了系统的整体吞吐量。将计算移动到数据附近，比将数据移动到应用附近显然更好。HDFS 为应用提供了将计算移动到数据附近的接口。

2.1.2 HDFS 架构

HDFS 采用了典型的 master/slave 架构设计。一个 HDFS 集群包含了一个活动的 NameNode，它的职责是管理文件系统名字空间以及处理用户对数据的访问。另外，集群中还包含了一定数量的 DataNode。一般是一个节点对应一个 DataNode，每一个 DataNode 都管理并存储了整体数据的一部分。HDFS 对外暴露了文件系统的名字空间，用户能够以文件的形式在上面存储数据。从内部看，一个文件通常被分成一个或多个 Block，这些 Block 存储在一组 DataNode 上。NameNode 会执行类似于打开、关闭或重命名文件夹或文件的操作，同时也负责确定数据块到具体 DataNode 的映射。DataNode 负责处理文件系统客户端的读写请求，在 NameNode 的统一调度下进行数据块的创建、删除和复制。HDFS 架构如图 2.1 所示。

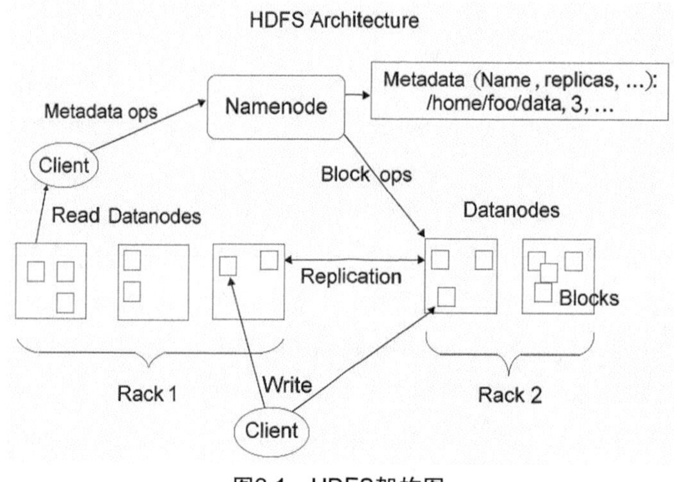

图2.1 HDFS架构图

1. **数据块**（Block）

HDFS 最基本的存储单位是数据块（Block），CDH 发行版默认的块大小（Block Size）是 128MB。

HDFS 中的文件被分成以 Block Size 为单位的数据块结构存储，小于一个块大小的文件不会占据整个块的空间。

2. **元数据节点**（NameNode）

NameNode 是管理者，一个 Hadoop 集群中只有一个活动的 NameNode 节点，它负责管理文件系统的命名空间和控制用户的访问。

NameNode 的主要功能如下。

➢ NameNode 提供名称查询服务。

➢ NameNode 保存元数据（metadata）信息。具体包括：文件包含哪些块，块保存在哪个 DataNode。元数据信息在 NameNode 启动后会加载到内存。

3. **数据节点**（DataNode）

一般而言，DataNode 是一个在 HDFS 实例中的单独机器上运行的进程。Hadoop 集群包含大量的 DataNode。DataNode 是文件系统中真正存储数据的地方，一个文件被拆分成多个块后，会将这些块存储在对应的 DataNode 上。客户端向 NameNode 发起请求，然后到对应的 DataNode 上写入或者读出对应的数据块。

DataNode 的主要功能如下。

（1）保存块，每个块对应一个元数据信息文件。这个文件主要描述块属于哪个文件、是文件中第几个块等信息。

（2）启动 DataNode 进程的时候向 NameNode 汇报块信息。

（3）通过向 NameNode 发送心跳与其保持联系（3 秒一次），如果 NameNode 在 10 分钟还没有收到 DataNode 的心跳，则认为该 DataNode 已经丢失，NameNode 会将该 DataNode 上的块复制到其他 DataNode。

4. **从元数据节点**（Secondary NameNode）

Secondary NameNode 并不是 NameNode 宕机时的备用节点，它的主要功能是周期性地将 EditLog 文件中对 HDFS 的操作合并到一个 FsImage 文件中，然后清空 Editlog 文件，防止日志文件过大。合并后的 FsImage 文件也在元数据节点保存了一份，NameNode 重启时就会加载最新的 FsImage 文件，这样周期性地合并可以减少 HDFS 重启的时间。Secondary NameNode 是用来帮助 NameNode 将内存中的元数据信息持久化到硬盘上的。

任务 2　使用 HDFS 处理移动通信数据文件

【任务描述】

本任务使用 HDFS shell 和 Java API 实现 HDFS 管理移动通信数据文件。

【关键步骤】

（1）使用 HDFS shell 完成移动通信业务数据管理操作。

（2）使用 Java API 完成移动通信业务数据管理操作。

2.2.1　使用 HDFS shell 操作完成移动通信数据的管理

HDFS 为用户提供了 shell 操作命令来管理 HDFS 上的数据。这些 shell 命令和 Linux 命令十分类似，已经熟悉 Linux 的用户可以更快速地对 HDFS 的数据进行操作。

HDFS 的基本命令格式如下。

【命令】

```
/bin/hdfs dfs -cmd <args>
```

注意

cmd 是具体的命令，cmd 前面的“-”不能省略。

1. 创建存放移动通信数据文件的目录

（1）列出目录结构

【命令】

hdfs dfs –ls 目录路径

示例 1

查看 HDFS 根目录下的文件。

```
[hadoop@hadoop hadoop]$ hdfs dfs -ls /
Found 1 items
drwxr-xr-x   - hadoop supergroup          0 2018-06-18 19:08 /kgc
```

（2）创建文件夹

【命令】

hdfs dfs –mkdir 文件夹名称

示例 2

在根目录下创建 HDFSShell 目录，用来存放移动通信数据文件。

```
[hadoop@hadoop hadoop]$ hdfs dfs -mkdir /HDFSShell
```

2. 将移动通信数据上传到 HDFS 并进行管理

（1）上传文件

【命令】

hdfs dfs –put|copyFromLocal|moveFromLocal 本地路径 HDFS 存放路径

示例 3

将本地 Linux 文件系统目录/home/hadoop/data/下的 mobile.txt 文件上传至示例 2 创建的 HDFSShell 目录下。

```
[hadoop@hadoop hadoop]$ hdfs dfs -put /home/hadoop/data/mobile.txt /HDFSShell
[hadoop@hadoop hadoop]$ hdfs dfs -ls /HDFSShell
Found 1 items
-rw-r--r--   1 hadoop supergroup       2833 2018-06-19 14:12 /HDFSShell/mobile.txt
```

注意

put、copyFromLocal、moveFromLocal 三个命令都可以用来上传文件，这里只演示了 put 命令的用法，其他两个命令请读者自己动手实践。

（2）查看文件

【命令】

hdfs dfs –text|cat HDFS 上的文件存放路径

示例 4

查看示例 3 上传的移动通信数据文件内容。

```
[hadoop@hadoop hadoop]$ hdfs dfs -text /HDFSShell/mobile.txt
13726230503    00-FD-07-A4-72-B8:CMCC 120.196.100.82    34523
13826544101    5C-0E-8B-C7-F1-E0:CMCC 120.197.40.4      234
13926435656    20-10-7A-28-CC-0A:CMCC 120.196.100.99    345
13926251106    5C-0E-8B-8B-B1-50:CMCC 120.197.40.4      2345
18211575961    94-71-AC-CD-E6-18:CMCC-EASY     120.196.100.99  2345
15684138413    5C-0E-8B-8C-E8-20:7DaysInn    120.197.40.4      234
15584138413    5C-0E-8B-8C-E8-20:7DaysInn    120.197.40.4 65345
15584138413    5C-0E-8B-8C-E8-20:7DaysInn    120.197.40.4 456
13560439658    C4-17-FE-BA-DE-D9:CMCC        120.196.100.99   345
15920133257    5C-0E-8B-C7-BA-20:CMCC        120.197.40.4      78
13719199419    68-A1-B7-03-07-B1:CMCC-EASY     120.196.100.82   2435
13660577991    5C-0E-8B-92-5C-20:CMCC-EASY     120.197.40.4 4534
15013685858    5C-0E-8B-C7-F7-90:CMCC 120.197.40.4 452
15989002119    E8-99-C4-4E-93-E0:CMCC-EASY     120.196.100.99   23452
```

注意

text、cat 两个命令都可以用来查看文件内容，这里只演示了 text 命令的用法，cat 命令请读者自己动手实践。

（3）下载文件

【命令】

hdfs dfs -get|copyToLocal|moveToLocal　HDFS 文件路径　本地存放路径

示例 5

将示例 3 上传的移动通信数据文件下载到本地用户的目录下。

```
[hadoop@hadoop hadoop]$ hdfs dfs -get /HDFSShell/mobile.txt /home/hadoop
```

注意

get、copyToLocal、moveToLocal 三个命令都可以用来下载文件，这里只演示了 get 命令的用法，其他两个命令请读者自己动手实践。

（4）统计目录下文件大小

【命令】

hdfs dfs -du　目录路径

示例 6

查看示例 2 创建的 HDFSShell 目录下各个文件的大小。

```
[hadoop@hadoop hadoop]$ hdfs dfs -du /HDFSShell
2833    2833    /HDFSShell/mobile.txt
```

（5）删除文件

【命令】

hdfs fs -rm|rmr 文件存放路径

示例 7

删除示例 3 上传的移动通信数据文件。

```
[hadoop@hadoop hadoop]$ hdfs dfs -rmr /HDFSShell/mobile.txt
```

在 HDFS 中删除文件与删除目录所使用的命令有所区别，一个是-rm，表示删除指定的文件或者空目录；一个是-rmr，表示递归删除指定目录及所有子目录和文件。如下所示的命令是删除 input 及所有子目录和文件。

```
hdfs dfs –rmr /home/hadoop/data/input
```

注意

在生产环境中要慎用-rmr，因为容易引起误删除操作。

3．HDFS shell 帮助命令

在 HDFS shell 使用过程中，可以使用 help 命令寻求帮助。

【命令】

hdfs dfs -help 命令

示例 8

查看 rm 命令的帮助。

```
[hadoop@hadoop hadoop]$ hdfs dfs -help rm
U-rm [-f] [-r|-R] [-skipTrash] <src> ... :
    Delete all files that match the specified file pattern. Equivalent to the Unix
    command "rm <src>"

    -skipTrash   option bypasses trash, if enabled, and immediately deletes <src>
    -f           If the file does not exist, do not display a diagnostic message or
                   modify the exit status to reflect an error.
    -[rR]        Recursively deletes directories
```

以上命令是日常工作中使用频次较高的，务必要熟练使用。如果需要学习其他 shell 命令，可以访问官网或使用 help 命令寻求帮助，在此不再赘述。了解 HDFS shell 的扩展内容请扫描二维码。

HDFS shell 命令更多操作

2.2.2 使用 Java API 操作完成移动通信数据的管理

HDFS 不仅提供了使用 HDFS shell 的方式来访问 HDFS 上的数据，还提供了以 Java API 的方式来操作 HDFS 上的数据。在实际开发中，大数据应用都是以代码的方式提交的，所以有必要掌握在代码中使用 API 的方式来操作 HDFS 数据。下面介绍如何使用 Java API 对 HDFS 中的文件进行操作。

1. 使用 IntelliJ IDEA 开发工具搭建开发环境

（1）在 IDEA 中创建 maven 项目 hdfsApp。

（2）添加 Maven 依赖包。Maven pom 文件如下。

```
Maven pom 文件
<properties>
    <project.build.sourceEncoding>UTF-8</project.build.sourceEncoding>
    <hadoop.version>2.6.0-cdh5.14.2</hadoop.version>
</properties>

<dependencies>
    <dependency>
        <groupId>org.apache.hadoop</groupId>
        <artifactId>hadoop-common</artifactId>
        <version>${hadoop.version}</version>
    </dependency>
    <dependency>
        <groupId>org.apache.hadoop</groupId>
        <artifactId>hadoop-hdfs</artifactId>
        <version>${hadoop.version}</version>
    </dependency>
    <dependency>
        <groupId>org.apache.hadoop</groupId>
        <artifactId>hadoop-client</artifactId>
        <version>${hadoop.version}</version>
    </dependency>
    <dependency>
        <groupId>junit</groupId>
        <artifactId>junit</artifactId>
        <version>4.11</version>
    </dependency>
</dependencies>
```

注意

添加的 HDFS 依赖包的版本要与实际环境安装的版本相同。

2. 使用 Java API 操作 HDFS

下面采用单元测试的方式来演示 Java API 操作 HDFS。在单元测试中，一般将初始化的操作放在 setUp 方法中完成，将关闭资源的操作放在 tearDown 方法中完成。

HDFS API 提供了访问文件系统的抽象父类 FileSystem，通过 FileSystem 类可以访问 HDFS 中的文件。在编写 HDFS 测试代码时，初始化 FileSystem 的操作可以放在 setUp 中，而关闭 FileSystem 的操作可以放在 tearDown 中。

注意

在执行单元测试之前，需要在$HADOOP_HOME/etc/hadoop/hdfs-site.xml 配置文件中添加如下配置，来关闭 HDFS 的权限检查。设置完成后，重启 HDFS 集群。

```
<property>
    <name>dfs.permissions</name>
    <value>false</value>
</property>
```

示例 9

创建单元测试的 setUp 和 tearDown 方法。

分析：

① 初始化 FileSystem 的基本方法是通过 FileSystem 类的静态方法 get(final URI uri,final Configuration conf,String user)获取一个实例。由于在 get 方法中使用到了 Configuration 配置对象，所以在初始化 FileSystem 对象前，需要先对 Configuration 对象进行初始化。

② 在 tearDown 方法中，需要将 Configuration 对象和 FileSystem 对象关闭。

关键代码：

```
package com.bigdata.hadoop.hdfs.api;
import org.apache.hadoop.fs.FileSystem;
import org.apache.hadoop.conf.Configuration;
import org.apache.hadoop.fs.Path;
import org.apache.hadoop.fs.BlockLocation;
import org.apache.hadoop.fs.FileStatus;
import org.apache.hadoop.fs.FSDataOutputStream;
import org.junit.After;
import org.junit.Before;
import org.junit.Test;
import util.HDFSUtil;
public class HDFSApp {
    public static final String HDFS_PATH = "hdfs://hadoop:8020/";
    FileSystem fileSystem = null;
    Configuration configuration = null;
    @Before
    public void setUp() throws Exception{
        configuration = new Configuration();
        fileSystem = FileSystem.get(new URI(HDFS_PATH), configuration,"hadoop");
    }
    @After
    public void tearDown() throws Exception{
        fileSystem = null;
        configuration = null;
```

```
        System.out.println("HDFSApp.tearDown()");
    }
}
```

初始化完成之后，就可以使用创建的 FileSystem 对象来完成管理移动通信数据文件的操作了。

（1）使用 FileSystem 对象的 mkdirs()方法创建文件夹

示例 10

在根目录下创建 HDFSJava 目录，用来存放移动通信数据文件。

关键代码：

```
/**
*创建/HDFSJava 目录
*/
@Test
public void create() throws Exception{
    fileSystem.mkdirs("/HDFSJava");
}
```

（2）使用 FileSystem.create()方法创建文件

示例 11

在示例 10 创建的目录下创建文件 mobiles.txt。

关键代码：

```
/**
*创建 mobiles.txt 文件
*/
@Test
public void createFile() throws Exception{
    FSDataOutputStream output = fileSystem.create (new Path(""/HDFSJava/mobiles.txt""));
    output.write("my telephone is *****".getBytes());
    output.flush();
    output.close();
}
```

（3）使用 FileSystem 对象的 rename()方法对文件进行重命名

示例 12

将 mobiles.txt 文件重命名为 newMobiles.txt。

关键代码：

```
/**
*重命名 mobiles.txt 文件
*/
@Test
public void rename() throws Exception{
    Path oldPath = new path("/HDFSJava/mobiles.txt");
    Path newPath = new Path("/HDFSJava/newMobiles.txt");
```

```
        System.out.println(fileSystem.rename(oldPath,newPath));
}
```

（4）使用 FileSystem 对象的 copyFromLocalFile()方法上传本地文件到 HDFS

示例 13

将 Linux 本地文件系统中的 mobiles.txt 文件上传到示例 10 创建的目录下。

关键代码：

```
/**
*上传 mobiles.txt 文件到指定的目录
*/
@Test
public void copyFromLocalFile() throws Exception{
      Path srcPath = new Path("/mobile/mobiles.txt");
      Path destPath = new Path("/HDFSJava");}
      fileSystem.copyFromLocalFile(srcPath,destPath);
}
```

（5）使用 FileSystem 对象的 listStatus()方法查看目录下的文件

示例 14

查看示例 10 创建的目录/HDFSJava 下的所有文件。

关键代码：

```
/**
*查看某个目录下的所有文件
*/
@Test
public static void listFiles() throws IOException {
       Path path = new Path("/HDFSJava");
       FileStatus[] list = fileSystem.listStatus(path);
             for (FileStatus f : list) {
             System.out.printf("name: %s, folder: %s, size: %d\n", f.getPath(),f.isDir(), f.getLen());
      }
}
```

（6）使用 FileSystem 对象的 getFileBlockLocations()方法查看文件块信息

示例 15

查看示例 13 上传的移动通信数据文件的文件块信息。

关键代码：

```
/**
*查看文件块信息
*/

@Test
public void getFileBlockLocations() throws Exception{
    FileStatus fileStatus = fileSystem.getFileStatus(new Path(HDFS_PATH +
                                    "HDFSShell/mobiles.txt"));
```

```
    BlockLocation[] blockLocations = fileSystem.getFileBlockLocations(fileStatus,
                0, fileStatus.getLen());
    for (BlockLocation block : blockLocations) {
        System.out.println(Arrays.toString(block.getHosts()) + "\t"
                + Arrays.toString(block.getNames()));
    }
}
```

FileSystem 对象还提供了其他一些方法来操作 HDFS，读者可以自行练习掌握。了解 FileSystem 更多扩展内容请扫描二维码。

FileSystem 更多操作

2.2.3　技能实训

使用 HDFS shell 命令完成对企业员工信息文件的操作。员工信息文件为“.txt”格式，员工信息包含的字段有姓名、年龄、性别、部门名称、学历、工作年限。参考数据格式如下：

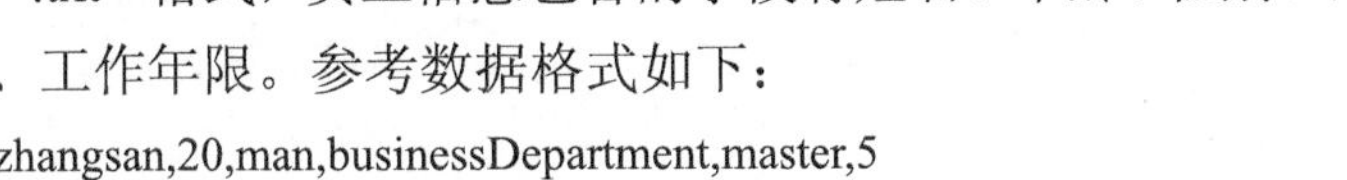

zhangsan,20,man,businessDepartment,master,5

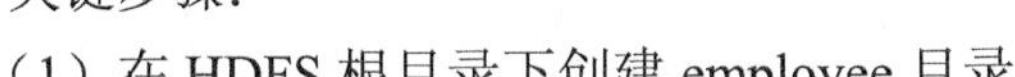

关键步骤：

（1）在 HDFS 根目录下创建 employee 目录。

（2）将 Linux 本地文件 employee.txt 上传到创建的 employee 目录下。

（3）使用 shell 命令查看上传的员工信息文件的内容。

任务 3　了解 HDFS 运行原理

【任务描述】

本任务使用 HDFS shell 和 Java API 实现 HDFS 管理移动通信数据文件。

【关键步骤】

（1）掌握 HDFS 读写执行流程。

（2）理解 HDFS 副本存放策略。

（3）了解 HDFS 文件数据的负载均衡和机架感知原理。

2.3.1　HDFS 读写流程

1．HDFS 文件读流程

客户端读取数据的过程如下。

（1）使用 HDFS 提供的客户端向远程的 NameNode 发起 RPC 请求。

（2）NameNode 会视情况返回文件的部分或者全部块列表，对于每个块，NameNode 都会返回该块副本的 DataNode 地址。

（3）客户端会选择离其最近的 DataNode 来读取块，如果客户端本身就是 DataNode，那么将从本地直接获取数据。

（4）读取完当前块数据后，关闭当前的 DataNode 连接，并为读取下一个块寻找最佳的 DataNode。每读取完一个块都会进行 checksum 验证，如果读取 DataNode 时出现错误，客户端会通知 NameNode，再从下一个拥有该块副本的 DataNode 继续读取。

（5）当读取完块列表后，并且文件读取还没有结束，客户端会继续向 NameNode 获取下一批块列表。

（6）一旦客户端完成读取操作后，就会调用 close 方法来完成资源的关闭操作。

HDFS 读数据流程如图 2.2 所示。

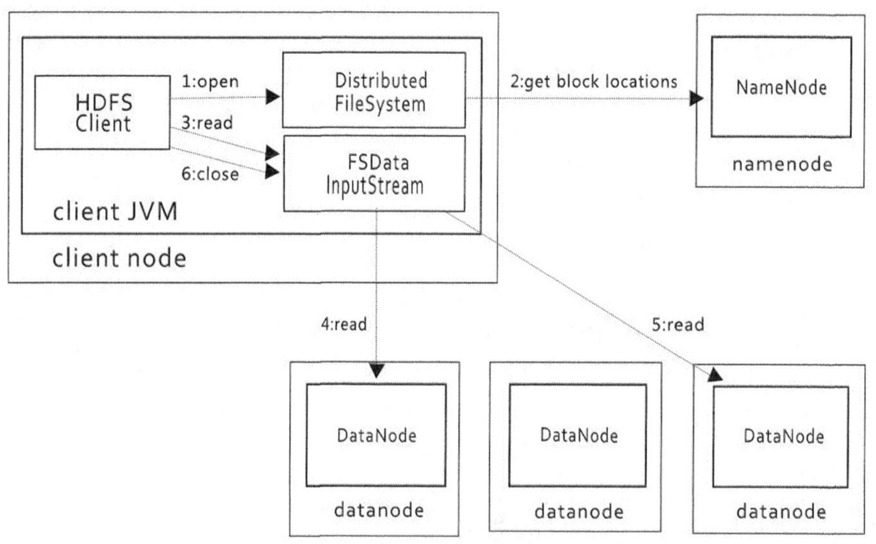

图2.2 HDFS读数据流程

2. HDFS 文件写流程

客户端写数据的过程如下。

（1）使用 HDFS 提供的客户端向远程的 NameNode 发起 RPC 请求。NameNode 会检查要创建的文件是否已经存在，创建者是否有权限进行操作，成功则为文件创建一个记录，否则会让客户端抛出异常。

（2）开始写文件时，客户端会将文件切分成多个 packets（数据包），并向 NameNode 申请块，获取用来存储副本的合适的 DataNode 列表。

（3）客户端调用 FSDataOutputStream API 的 write 方法，首先将其中一个块写到 DataNode 上，每一个块默认都有 3 个副本，并不是由客户端分别往 3 个 DataNode 上写 3 份数据，而是由已经上传了块的 DataNode 产生新的线程，由这个 DataNode 按照副本规则往其他 DataNode 写副本。

（4）FSDataOutputStream 内部维护着一个确认队列，当接收到所有 DataNode 确认写完的消息后，数据才会从确认队列中删除。

（5）客户端完成数据的写入后，会对数据流调用 close 方法来关闭相关资源。

HDFS 写数据流程如图 2.3 所示。

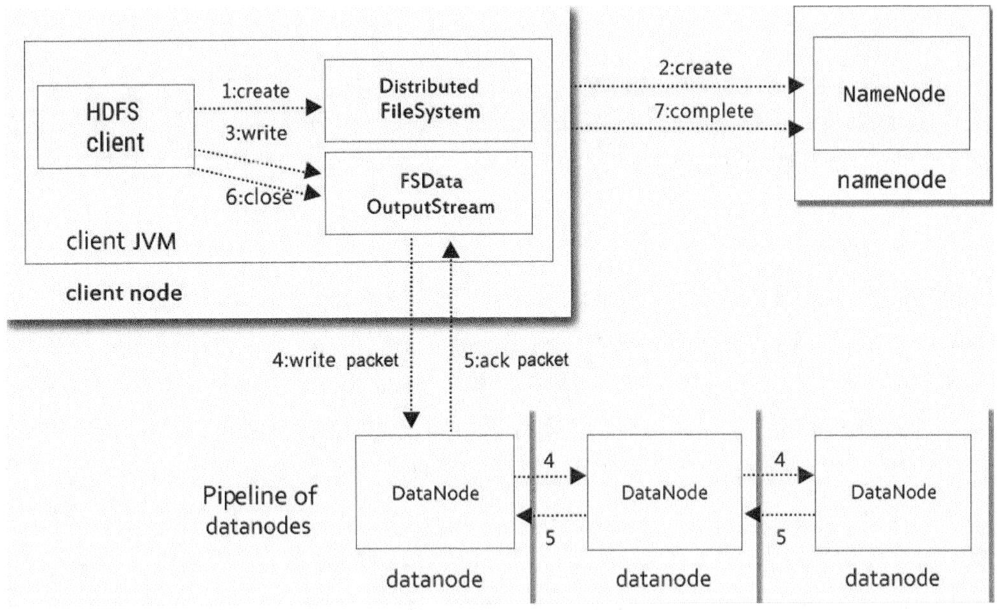

图2.3　HDFS写数据流程

2.3.2　HDFS 副本机制

副本的存放位置会严重影响 HDFS 的可靠性和性能，HDFS 上的文件对应的块保存有多个副本且提供容错机制，因此副本丢失或宕机时能够自动恢复，默认保存 3 个副本。HDFS 副本摆放机制如图 2.4 所示。

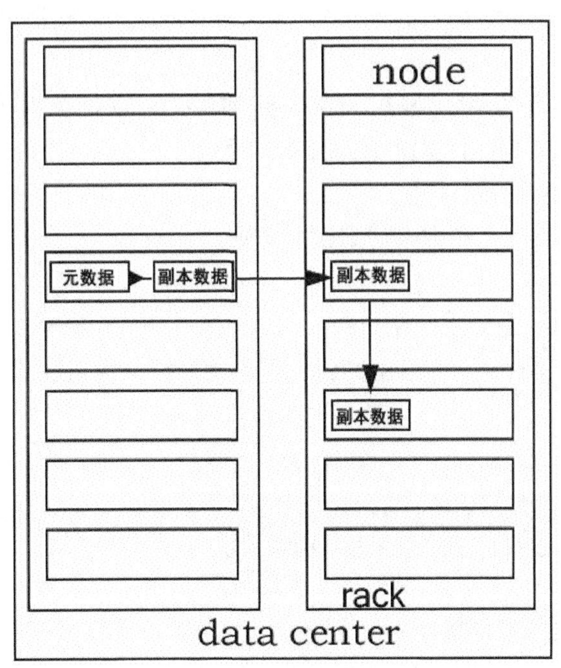

图2.4　HDFS副本摆放机制

1. 副本摆放策略

一般情况下，当副本设置为 3 时，HDFS 的副本摆放策略是将第一个副本放置在上传文件的 DataNode 上，第二个副本放置在同一机架的不同节点上，第三个副本放置在不同机架上。该策略可以减少机架间的写流量，提高写性能，并且该策略不会影响数据的可靠性和可用性。

2. 副本系数

上传文件到 HDFS 时，可以通过设置客户端属性 dfs.replication 来指定上传文件的副本系数，一旦设置，文件的块副本数就确定了，无论以后怎么更改副本系数，这个文件的副本数都不会改变。如果客户端未指定具体的 replication，将会去配置文件中读取，采用默认的副本系数；文件上传后，副本数已经确定，再修改 dfs.replication 既不会影响以前的文件，也不会影响后面指定副本数的文件，只会影响后面采用默认副本数的文件。

在实际开发中，如果在 hdfs-site.xml 中设置了 dfs.replication=1，并不意味着块的副本数就是 1，如果没把 hdfs-site.xml 加入到工程的 classpath 里，那么程序运行时读取的就是 hdfs-default.xml 中 dfs.replication 的默认值 3，这也是 dfs.replication 总是 3 的原因。

2.3.3 HDFS 负载均衡

HDFS 的集群架构支持数据均衡策略。HDFS 集群可能会出现机器与机器之间磁盘利用率不平衡的情况，例如：集群内新增、删除节点，或者某个节点机器的硬盘存储达到饱和。当数据不均衡时，会导致网络带宽的消耗，将无法很好地进行本地计算。当 HDFS 负载不均衡时，需要对 HDFS 进行数据的负载均衡调整，即对各节点机器上数据的存储分布进行调整，从而让数据均匀地分布在各个 DataNode 上，均衡 IO 性能，防止热点的发生。

在 Hadoop 中，通过运行 start-balance.sh 脚本，可以启动 HDFS 的数据均衡服务 Balancer。该脚本存放在$HADOOP_HOME/bin 目录下，启动命令为：$HADOOP_HOME/bin/start-balancer.sh -threshold。

用户可以通过设置以下参数来达到数据均衡。

（1）-threshold

默认设置：10，参数取值范围：0～100。

参数含义：判断集群是否平衡的阈值。理论上，该参数值设置得越小，整个集群越平衡。

（2）dfs.datanode.balance.bandwidthPerSec

默认设置：1048576（1MB/s）

参数含义：Balancer 运行时允许占用的带宽。

2.3.4 HDFS 机架感知

通常大型 Hadoop 集群是以机架的形式来组织的，同一个机架上不同节点间的网络状况比不同机架之间的网络状况更为理想，NameNode 设法将数据块副本保存在不同的机架上以提高容错性。

Hadoop 对机架的感知并非是自适应的，即 Hadoop 集群需要手动指定机器是属于哪个机架而非智能地感知。Hadoop 允许集群管理员通过配置 dfs.network.script 参数来确定节点所处的机架，配置文件提供了 ip 到机架（rackid）的翻译。NameNode 通过这个配置可以知道集群中各个 DataNode 机器的 rackid，如果 topology.script.file.name 没有设定，则每个 ip 都会被翻译成/default-rack。

当没有配置机架信息时，所有的机器都在同一个默认的机架下，名为“/default-rack”。这种情况下，任何一台 DataNode 机器，不管物理上是否属于同一个机架，都会被认为是在同一个机架下。

一旦配置了 topology.script.file.name，就可以按照网络拓扑结构来寻找 DataNode；topology.script.file.name 这个配置选项的 value 则指定为一个可执行程序，通常为一个脚本，该脚本接受一个参数，输出一个值。接受的参数通常为某台 DataNode 机器的地址，而输出的值通常为该地址对应的 DataNode 所在的地址。

任务 4 实现移动通信数据的行文件方式存储

【任务描述】

了解 HDFS 的序列化机制和常用文件格式，掌握 SequenceFile 和 MapFile 的常用操作。

【关键步骤】

（1）了解 Hadoop 序列化机制。

（2）了解 Hadoop 的文件格式。

2.4.1 Hadoop 序列化机制

1. 序列化和反序列化的定义

在分布式数据处理中，序列化和反序列化主要应用于进程间通信和永久存储两个领域。

序列化（Serialization）是指将结构化对象转化为字节流，这样便于在网络上传输或者写在磁盘上进行永久存储。

反序列化（Unserialization）是指将字节流转回成结构化对象的逆过程。

在 Hadoop 系统中，多个节点间通信是通过远程过程调用（Remote Procedure Call，RPC）实现的，RPC 协议将消息序列化为二进制流后发送到远程节点，远程节点再将二进制流反序列化为消息。

2. Hadoop 文件的序列化

Hadoop 文件的序列化并没有采用 Java 的序列化机制，而是自己开发了一套文件的序列化框架。在 Hadoop 的序列化框架中，对象可以复用，因此减少了对象的分配和回收所需的资源，极大地提高了应用效率。

在 Hadoop 系统中，通过 Writable 或 WritableComparable 接口可以实现序列化机制，WritableComparable 接口还提供了比较功能。Writable 接口和 WritableComparable 接口的定义如下。

```
public interface Writable{
    void write(DataOutput out) throws IOException;//状态写入到 DataOutput 二进制流
    void readFields(DataInput in) throws IOException;//从 DataInput 二进制流中读取状态
}
public interface WritableComparable<T> extends Writable,Comparable<T>{
}
```

序列化扩展

示例 16

编写 Mobile 类，使用 Hadoop 序列化的方式序列化 Mobile 类，并将序列化的对象反序列化出来。

实现步骤：

（1）编写 Mobile 实体类。

（2）编写序列化工具类，实现序列化和反序列化方法。

（3）编写测试类来测试序列化和反序列化方法。

关键代码：

```
import org.apache.hadoop.io.IntWritable;
import org.apache.hadoop.io.Text;
import org.apache.hadoop.io.WritableComparable;
import java.io.DataInput;
import java.io.DataOutput;
import java.io.IOException;
/**
*序列化 Mobile 类
*/
public class Mobile implements WritableComparable<Mobile> {
    private Text name = new Text();
    private IntWritable phone = new IntWritable();
    private Text sex = new Text();

    public Mobile(String name,int phone,String sex){
        this.name.set(name);
        this.phone.set(phone);
        this.sex.set(sex);
    }

    public Mobile(Text name,IntWritable phone,Text sex){
        this.name=name;
        this.phone=phone;
        this.sex=sex;
    }
```

```
public Mobile(){
}

public int compareTo(Mobile o) {
    int resoult = 0;
    int comp1 = name.compareTo(o.name);
    if(comp1!=0){
        return comp1;
    }
    int comp2 = phone.compareTo(o.phone);
    if(comp2!=0){
        return comp2;
    }
    int comp3 = sex.compareTo(o.sex);
    if(comp3!=0){
        return comp3;
    }
    return resoult;
}

public void write(DataOutput out) throws IOException {
    name.write(out);
    phone.write(out);
    sex.write(out);
}

public void readFields(DataInput in) throws IOException {
    name.readFields(in);
    phone.readFields(in);
    sex.readFields(in);
}

@Override
public int hashCode() {
    final int prime = 31;
    int result = 1;
    result = prime * result + ((phone==null)?0:phone.hashCode());
    result = prime * result + ((name==null)?0:name.hashCode());
    result = prime * result + ((sex==null)?0:sex.hashCode());
    return result;
}

@Override
public boolean equals(Object obj) {
```

```
            if(this==obj)
                return   true;
            if(obj==null)
                return false;
            if(getClass()!=obj.getClass())
                return false;
            Mobile mobile = (Mobile)obj;
            if(phone==null) {
                if (mobile.phone != null)
                    return false;
            }else if(!phone.equals(mobile.phone))
                return false;

            if(name==null){
                if(mobile.name!=null)
                    return false;
            }else if(!name.equals(mobile.name)){
                return false;
            }

            if(sex==null){
                if(mobile.sex!=null)
                    return false;
            }else if(!sex.equals(mobile.sex)){
                return false;
            }
            return true;
        }

        @Override
        public String toString() {
            return "Mobile [name="+name+",phone="+phone+",sex="+sex+"]";
        }
}
```

序列化工具类：

```
package com.bigdata.hadoop.hdfs.io;
import org.apache.hadoop.io.Writable;
import java.io.*;
/**
*序列化操作
*/
public class HadoopSerializationUtil {
    public static byte[] serialize(Writable writable) throws IOException{
        ByteArrayOutputStream out = new ByteArrayOutputStream();
        DataOutputStream dataout = new DataOutputStream(out);
```

```
            writable.write(dataout);
            dataout.close();
            return out.toByteArray();
        }

        public static void deserialize(Writable writable,byte[] bytes) throws Exception{
            ByteArrayInputStream in = new ByteArrayInputStream(bytes);
            DataInputStream datain = new DataInputStream(in);
            writable.readFields(datain);
            datain.close();
        }
}
```

测试类：

```
import util.HadoopSerializationUtil;
public class mobileTest {
    public static void main(String[] args) throws Exception{
        //测试序列化
        Mobile mobile = new Mobile("lisi", 13519562548, "man");
        byte[] values = HadoopSerializationUtil.serialize(mobile);

        //测试反序列化
        Mobile m = new Mobile();
        HadoopSerializationUtil.deserialize(m, values);
        System.out.println(m);
    }
}
```

输出结果：

```
Mobile[name=lisi,phone=13519562548,sex=man]
```

2.4.2　文件格式

1．简介

Hadoop 中支持的文件格式可以分为面向行和面向列两类。

面向行是指同一行的数据存储在一起，即连续存储，SequenceFile、MapFile、Avro、DataFile 都采用行存储方式。即便只访问同行的小部分数据，也需要将整行读入内存，所以行存储方式适合整行数据同时处理的情况。

面向列是指将整个文件切割成若干列数据，每列数据一起存储，Parquet、RCFile、ORCFile 都采用列存储方式。采用列存储方式，读取数据时可以跳过不需要的列，但是需要更多的存储空间，因为需要缓存行在内存中，而且一旦写入失败，文件将无法恢复。

本节主要针对面向行存储的 SequenceFile 和 MapFile 进行讲解，其他存储方式后面用到时再详细讲解。

2．SequenceFile

SequenceFile 是 Hadoop API 提供的一种对二进制文件的支持。SequenceFile 的存储类

似于 Log 文件，所不同的是 Log 文件的每条记录是纯文本数据，而 SequenceFile 的每条记录是序列化的字符数组。SequenceFile 用于解决大量小文件（所谓小文件，泛指小于块大小的文件）问题，可以直接将<key,value>对序列化到文件中，一般对小文件可以使用 SequenceFile 进行合并，即将文件名作为 key，文件内容作为 value，再序列化到大文件中。

SequenceFile 的优点：

- 支持压缩。可以定制基于 Record（记录）和 Block（块）进行压缩。
- 本地化任务支持。因为文件可以被切分，因此在运行 MapReduce 任务时数据的本地化情况应该是非常好的。

示例 17

创建 mobileSeqFile.seq 文件，并将移动通信数据写入到该文件中。

实现步骤：

① 设置 Configuration。

② 获取 FileSystem。

③ 设置文件输出路径。

④ 利用 SequenceFile.createWriter()创建 SequenceFile.Writer。

⑤ 调用 SequenceFile.Write.append 追加写入。

⑥ 关闭流。

关键代码：

```
import org.apache.hadoop.conf.Configuration;
import org.apache.hadoop.fs.FileSystem;
import org.apache.hadoop.fs.Path;
import org.apache.hadoop.io.IOUtils;
import org.apache.hadoop.io.IntWritable;
import org.apache.hadoop.io.SequenceFile;
import org.apache.hadoop.io.Text;
import java.io.IOException;
import java.net.URI;
import java.net.URISyntaxException;
/**
*SequenceFile 写操作
*/
public class SequenceFileWriter {
    private static Configuration configuration = new Configuration();
    private static String url = "hdfs://hadoop:8020/";
    private    static String[] data = {"lisi","13519562548","man"};
    public static void main(String[] args) throws Exception {
        Configuration conf = new Configuration();
        FileSystem fileSystem = null;
        fileSystem = FileSystem.get(new URI(url),conf,"hadoop");
        Path outputPath = new Path("MySequenceFile.seq");
        IntWritable key = new IntWritable();
```

```
            Text value = new Text();
            SequenceFile.Writer writer =
                SequenceFile.createWriter(fileSystem,configuration,outputPath,IntWritable.class,Text.class);
            for(int i=0;i<10;i++){
                key.set(10-i);
                value.set(data[i%data.length]);
                writer.append(key,value);
            }
            IOUtils.closeStream(writer);
        }
}
```

示例 18

从示例 17 创建的 mobileSeqFile.seq 文件中读取移动通信数据。

实现步骤：

① 设置 Configuration。

② 获取 FileSystem。

③ 设置文件输入路径。

④ 利用 SequenceFile.Reader()创建读取类 SequenceFile.Reader。

⑤ 获取 key 和 value 的 class。

⑥ 读取。

⑦ 关闭流。

关键代码：

```
import java.net.URI;
import org.apache.hadoop.conf.Configuration;
import org.apache.hadoop.fs.FileSystem;
import org.apache.hadoop.fs.Path;
import org.apache.hadoop.io.IOUtils;
import org.apache.hadoop.io.*;
import org.apache.hadoop.util.ReflectionUtils;
/**
*SequenceFile 读操作
*/
public class SequyenceFileReader {
    private static Configuration configuration = new Configuration();
    private static String url = "hdfs://hadoop:8020/";
    public static void main(String[] args) throws Exception {
        FileSystem fileSystem = null;
        fileSystem = FileSystem.get(new URI(url),configuration,"hadoop");
        Path inputPath = new Path("MySequenceFile.seq");
        SequenceFile.Reader reader = new SequenceFile.Reader(fileSystem,inputPath,configuration);
        Writable keyClass =(Writable) ReflectionUtils.newInstance(reader.getKeyClass(),
                    configuration);
```

```
            Writable valueClass =(Writable)   ReflectionUtils.newInstance(reader.getValueClass(),
                        configuration);
            while(reader.next(keyClass,valueClass)){
                System.out.println("key:"+keyClass);
                System.out.println("value:"+valueClass);
                System.out.println("position:"+reader.getPosition());
            }
            IOUtils.closeStream(reader);
        }
    }
```

3. MapFile

MapFile 是经排序的 SequenceFile，由两部分构成，分别是 data 和 index。index 作为文件的数据索引，主要记录了每个 Record 的 key 值，以及该 Record 在文件中的偏移位置。在 MapFile 被访问的时候，索引文件先被加载到内存，再通过 index 的映射关系迅速定位到指定 Record 所在文件中位置。MapFile 的检索效率更高，缺点是会消耗一部分内存来存储 index 数据。

示例 19

创建 mobileMapFile.map，并将移动通信数据写入到该文件中。

实现步骤：

① 设置 Configuration。

② 获取 FileSystem。

③ 设置文件输出路径。

④ 利用 MapFile.Writer()创建 MapFile.Writer。

⑤ 调用 MapFile.Writer.append 追加写入。

⑥ 关闭流。

关键代码：

```
import org.apache.hadoop.conf.Configuration;
import org.apache.hadoop.fs.FileSystem;
import org.apache.hadoop.fs.Path;
import org.apache.hadoop.io.IOUtils;
import org.apache.hadoop.io.MapFile;
import org.apache.hadoop.io.Text;
import java.net.URI;
/**
 * MapFile 写操作
 */
public class MapFileWriter {
    static Configuration configuration = new Configuration();
    private static String url = "hdfs://hadoop:8020";
    public static void main(String[] args) throws   Exception{
        FileSystem fileSystem = FileSystem.get(new URI(url),configuration,"hadoop");
        Path outPath = new Path("mobileMapFile.map");
```

```
            Text key = new Text();
            key.set("mobilemapkey");
            Text value = new Text();
            value.set("mobilemapvalue");
            MapFile.Writer writer =
                new MapFile.Writer(configuration,fileSystem,outPath.toString(),Text.class,Text.class);
            writer.append(key,value);
            IOUtils.closeStream(writer);
        }
}
```

示例 20

从示例 19 创建的 mobileMapFile.map 文件中读取移动通信数据。

实现步骤：

① 设置 Configuration。

② 获取 FileSystem。

③ 设置文件输入路径。

④ 利用 MapFile.Reader()创建读取类 MapFile.Reader。

⑤ 获取 key 与 value 的 class。

⑥ 读取。

⑦ 关闭流。

关键代码：

```
import org.apache.hadoop.conf.Configuration;
import org.apache.hadoop.fs.FileSystem;
import org.apache.hadoop.fs.Path;
import org.apache.hadoop.io.*;
import org.apache.hadoop.util.ReflectionUtils;
import java.net.URI;

/**
 * MapFile 读操作
 */
public class MapFileReader {
    static Configuration configuration = new Configuration();
    private static String url = "hdfs://hadoop:8020";
    public static void main(String[] args) throws   Exception{
        FileSystem fileSystem = FileSystem.get(new URI(url),configuration,"hadoop");
        Path inPath = new Path("mobileMapFile.map");
        MapFile.Reader reader = new MapFile.Reader(fileSystem,inPath.toString(),configuration);
        Writable keyclass =(Writable) ReflectionUtils.newInstance(reader.getKeyClass(),
                configuration);
        Writable valueclass =(Writable) ReflectionUtils.newInstance(reader.getValueClass(),
                configuration);
```

```
            while (reader.next((WritableComparable)keyclass,valueclass)){
                System.out.println(keyclass);
                System.out.println(valueclass);
            }
            IOUtils.closeStream(reader);
        }
    }
```

2.4.3 技能实训

使用 Java API 创建 mySequenceFile.seq 文件，将个人信息写入到该文件中。其中，个人信息包含的字段有姓名（name）、年龄（age）、性别（sex）、家庭住址（address）、学历（education）。

具体实现步骤如下。

（1）设置 Configuration。

（2）获取 FileSystem。

（3）设置文件输出路径。

（4）利用 SequenceFile.createWriter()创建 SequenceFile.Writer。

（5）调用 SequenceFile.Write.append 追加写入。

（6）关闭流。

本章总结

- HDFS 是 Hadoop 的核心构成，可以部署在普通的设备上，是企业大数据存储的首选方案。
- 在 HDFS 的体系架构中，NameNode 负责管理名称空间和元数据信息，DataNode 负责存储数据块。
- 用户可以使用 HDFS shell 和 Java API 两种方式访问 HDFS。
- HDFS 支持行文件存储和列文件存储格式。

本章作业

一、简答题

1．HDFS 核心组件有哪些？每个组件的具体功能是什么？

2．为什么 HDFS 不适合存储小文件？

二、编码题

1．将本地文件夹下的日志小文件合并，并上传到 HDFS 根目录下的/mergeSmallFiles/result/目录下。

2．使用 shell 命令在根目录下新建 myHdfsFile.txt 文件。

3．使用 Java API 的方式获得 myHdfsFile.txt 文件的最后一次修改时间。

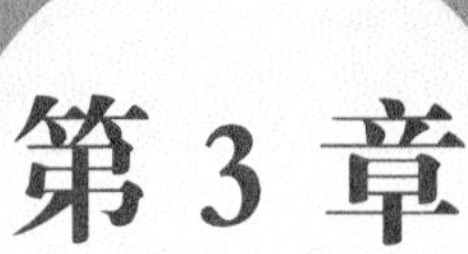

Hadoop 分布式计算框架 MapReduce

技能目标

- 了解 MapReduce
- 理解 MapReduce 编程模型
- 会使用 MapReduce 核心 API 编程
- 掌握 MapReduce 输入/输出格式

本章任务

任务 1　使用 MapReduce 完成词频统计功能
任务 2　按号段统计手机号码
任务 3　使用 MapReduce 编写应用案例

本章资源下载

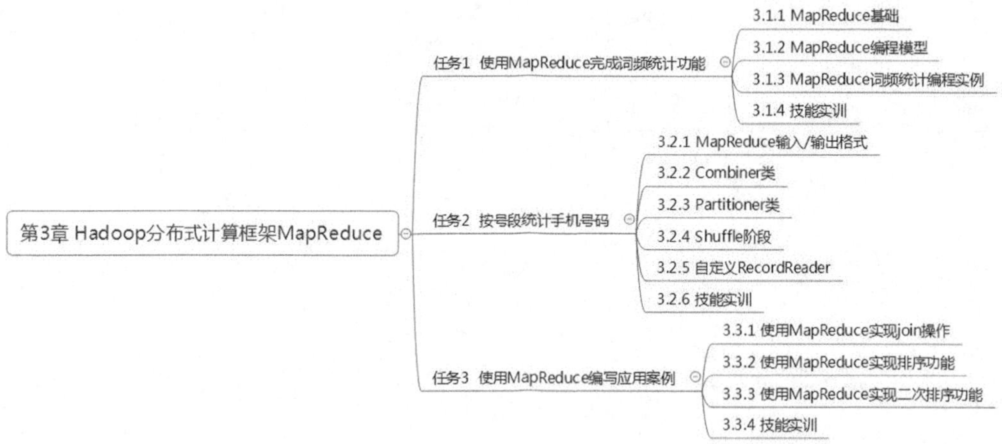

Hadoop 不仅提供了分布式存储的解决方案，同时提供了基于分布式文件系统 HDFS 进行分布式计算的解决方案，也就是本章要介绍的 MapReduce 框架。MapReduce 是 Hadoop 的一个核心构成。使用 MapReduce 框架，用户可以很方便地编写分布式应用程序。本章通过丰富的案例演示介绍如何使用 MapReduce 解决实际问题，同时对 MapReduce 的执行流程进行阐述。

任务 1 使用 MapReduce 完成词频统计功能

【任务描述】

使用 MapReduce 实现词频统计功能。

【关键步骤】

（1）了解 MapReduce。

（2）理解 MapReduce 编程模型。

（3）使用 MapReduce 实现词频统计功能。

3.1.1 MapReduce 基础

1. MapReduce 概述

MapReduce 是 Hadoop 中面向大数据并行处理的计算模型、框架和平台，其设计思想来源于 Google 公司发表的 MapReduce 论文。MapReduce 用于海量数据的并行计算，它采用“分而治之”的思想，把大规模数据集的操作分发到多个机器去共同完成，然后对各个节点的中间结果进行整合后得到最终的结果。基于 MapReduce 编写的应用可以以并行计算的方式在多个计算机节点上处理大量的数据。MapReduce 的设计目标是方便编程人员在不熟悉分布式并行开发的情况下，将编写的程序运行在分布式系统上，从而降低了分布式开发的入门门槛。

2. MapReduce 特点

MapReduce 作为一个用于处理海量数据的分布式计算框架，具有以下优点。

（1）易于编程。MapReduce 提供了方便用户开发的编程接口，开发人员只需要简单地实现接口，就可以完成一个分布式程序，并且这个分布式程序可以运行在大量廉价的机器上。

（2）平滑无缝的可扩展性。MapReduce 集群的构建完全选用价格便宜、易于扩展的廉价服务器，因此可以通过简单地增加机器来提高 MapReduce 集群的计算性能。多项研究表明，对于很多计算问题，基于 MapReduce 的计算性能可随着节点数目的增长保持近乎线性的增长。

（3）高容错性。MapReduce 集群中使用了大量的廉价服务器，因此节点硬件失效和软件出错是常态。但 MapReduce 框架使用了多种有效的错误检测和恢复机制，使集群具有应对节点失效的健壮性。比如其中一个节点失效，其他节点能够无缝接管失效节点的计算任务，当失效节点恢复后，又能够无缝加入集群。这个过程不需要人工进行配置，全部由 Hadoop 内部完成。

（4）高吞吐量。MapReduce 能对 PB 量级以上的海量数据进行离线处理，它利用集群中的大量数据存储节点同时访问数据，从而利用分布集群中大量节点上的磁盘集合提供高吞吐量的数据访问和传输。

MapReduce 虽然具有许多优势，但也有不擅长的领域。

（1）难以提供实时计算。MapReduce 处理的是存储在磁盘上的数据，受磁盘读写速度的限制，不能像 MySQL 和 NoSQL 数据库一样，实时地返回结果。

（2）不能流式计算。MapReduce 处理的是存储在磁盘上的静态数据，而流式计算的输入数据是动态的。

（3）难以用于 DAG（有向无环图）计算。DAG 计算中，由于多个任务间存在依赖关系，后一个应用的输入可能是前一个应用的输出。而 MapReduce 作业的输出结果都会写到磁盘上，对于存在复杂依赖关系的 DAG 计算，使用 MapReduce 会造成大量的磁盘 IO，降低集群的使用性能。

3.1.2 MapReduce 编程模型

1. 编程模型概述

MapReduce 是一种编程模型，其创意和灵感来源于函数式编程。MapReduce 从名称上就表现出它的核心原理，即由 Map 和 Reduce 两个阶段组成。Map 表示“映射”，由一定数量的 Map Task 组成。Reduce 表示“归约”，由一定数量的 Reduce Task 组成。用户只需要编写 map()和 reduce()函数，就可以完成分布式程序的设计。

在 Map 阶段，用户自定义的 map()函数接收<key,value>对作为输入，执行 map()函数逻辑，产生新的<key,value>对并写入本地磁盘，MapReduce 框架会对这些中间结果进行排序，并将 key 值相同的数据放在一起形成新的列表，统一交给 reduce()函数处理。

在 Reduce 阶段，reduce()函数接收 key 及对应的 value 列表作为输入，将 key 值相同的 value 值合并后，产生新的<key,value>对输出到 HDFS 上。

MapReduce 应用程序的执行过程如图 3.1 所示。关于 MapReduce 编程模型的详细介绍请扫描二维码。

MapReduce
编程模型

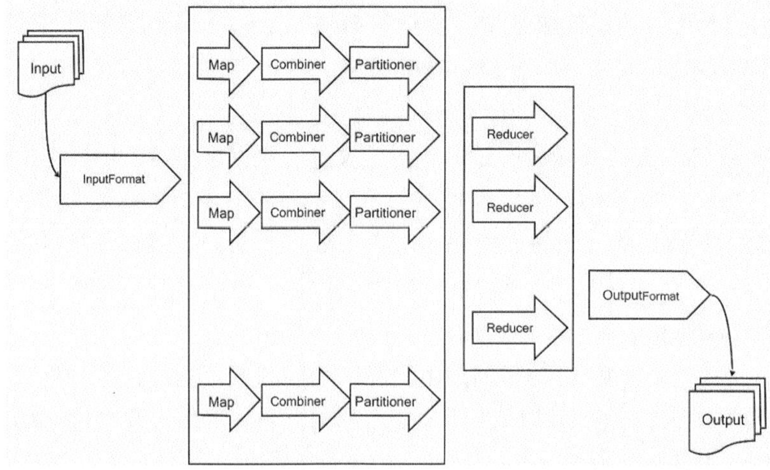

图3.1 MapReduce执行过程

2. MapReduce 编程三部曲

（1）输入 Input。MapReduce 输入一系列 k1/v1 对。

（2）Map 和 Reduce 阶段。Map：(k1,v1) -> list(k2,v2)，Reduce：(k2,list(v2)) -> list(k3,v3)。其中 k2/v2 是中间结果对。

（3）输出 Output。MapReduce 输出一系列 k3/v3 对。

3.1.3 MapReduce 词频统计编程实例

在理解了 MapReduce 的基本编程模型及原理后，接下来使用 MapReduce 编程完成词频统计（WordCount）功能。通过示例，进一步加深对 MapReduce 各部分执行流程的理解。

1. 需求分析

表 3-1 所示为 WordCount 的输入/输出内容。

表 3-1 WordCount 的输入/输出

输入	输出
Hello Hadoop BigData Hello Hadoop MapReduce Hello Hadoop HDFS BigData Perfect	BigData 2 Hadoop 3 HDFS 1 Hello 3 MapReduce 1 Perfect 1

开发 MapReduce 程序，一般包含三个部分：一个 map()函数、一个 reduce()函数和一个 main()函数。运行时，计算任务会分为 Map 阶段和 Reduce 阶段。

下面分析 Map 和 Reduce 阶段的输入/输出结果。

（1）Map 阶段。并行读取文本，对每个单词执行自定义的 map()函数，形成<key,value>对。具体过程如下。

读取第一行“Hello Hadoop BigData”，执行 map()函数后，分割单词形成 Map。

<Hello,1><Hadoop,1><BigData,1>

读取第二行“Hello Hadoop MapReduce”，执行 map()函数后，分割单词形成 Map。

<Hello,1><Hadoop,1><MapReduce,1>

读取第三行“Hello Hadoop HDFS”，执行 map()函数后，分割单词形成 Map。

<Hello,1><Hadoop,1><HDFS,1>

读取第四行“BigData Perfect”，执行 map()函数后，分割单词形成 Map。

<BigData,1><Perfect,1>

（2）Reduce 阶段。Reduce 操作是对 Map 的结果进行排序、合并，最后得出词频结果。reduce()函数执行前会经过 Shuffle（混排）等过程，将形成的 Map 中间结果中相同的 key 组合成 value 数组，结果如下。

<BigData,[1,1]><Hadoop,[1,1,1]><HDFS,1><Hello,[1,1,1]><MapReduce,[1]><Perfect,[1]>

然后 Reduce 端循环执行 Reduce(K,V[])，分别统计每个单词出现的次数，得到结果如下。

<BigData,2><Hadoop,3><HDFS,1><Hello,3><MapRdeuce,1><Perfect,1>

2. WordCount 代码实现

了解执行过程之后，接下来就要使用代码的方式来实现词频统计功能。本节选用的开发工具是 IntelliJ IDEA，使用 Maven 完成 jar 管理。

（1）新建 Maven 工程 mapreduceTest，添加 pom 文件依赖关系，代码如下。

```
<properties>
    <project.build.sourceEncoding>UTF-8</project.build.sourceEncoding>
    <hadoop.version>2.6.0-cdh5.14.2</hadoop.version>
</properties>
<dependencies>
    <dependency>
        <groupId>org.apache.hadoop</groupId>
        <artifactId>hadoop-common</artifactId>
        <version>${hadoop.version}</version>
    </dependency>
    <dependency>
        <groupId>org.apache.hadoop</groupId>
        <artifactId>hadoop-hdfs</artifactId>
        <version>${hadoop.version}</version>
    </dependency>
    <dependency>
        <groupId>org.apache.hadoop</groupId>
```

```
            <artifactId>hadoop-mapreduce-client-core</artifactId>
            <version>${hadoop.version}</version>
        </dependency>
    </dependencies>
```

（2）在 WordCount 类中编写 WordMapper 类。

MapReduce API 中提供了抽象类 Mapper<Object,Text,Text,IntWritable>，编写词频统计的 WordMapper 类需要继承该抽象类并实现如下方法。

```
public void map(Object key,Text value,Context context) throws IOException,InterruptedException
```

map 方法是 Mapper 抽象类的核心方法，它有三个参数。

Object key：通常是行偏移量，输入格式不同，含义也不同，表示输入的 key。

Text value：一行文本内容，表示输入的 value。

Context context：记录的是 Map 端的整个上下文，可以使用该对象将数据写到磁盘。

示例 1

编写实现词频统计功能的 WordMapper 类。

关键代码：

```
public static class WordMapper extends Mapper<Object,Text,Text,IntWritable>{
    public static final IntWritable one = new IntWritable(1);
    private Text word = new Text();
    public void map(Object key, Text value, Context context)
            throws IOException, InterruptedException {
        StringTokenizer itr = new StringTokenizer(value.toString());
        while (itr.hasMoreTokens()) {
            this.word.set(itr.nextToken());
            context.write(this.word, one);
        }
    }
}
```

（3）在 WordCount 类中编写 WordReducer 类。

MapReduce API 中提供了抽象类 Reducer<Text, IntWritable,Text,IntWritable>，编写词频统计的 WordReducer 类需要继承该抽象类并实现如下方法。

```
public void reduce(Text key,Iterable<IntWritable> values,Context context) throws IOException,
InterruptedException
```

reduce 方法是 Reducer 抽象类的核心方法，它有三个参数。

Text key：Map 端输出的 key 值。

Iterable<IntWritable> values：Map 端输出的 value 集合（相同 key 的集合）。

Context context：Reduce 端上下文，与 Map 端参数作用一致。

示例 2

编写实现词频统计功能的 WordReducer 类。

关键代码：

```
public static class WordReducer extends Reducer<Text, IntWritable, Text, IntWritable> {
```

```
    private IntWritable result = new IntWritable();
    public void reduce(Text key, Iterable<IntWritable> values, Context context) throws IOException,
                        InterruptedException {
        int sum = 0;
        IntWritable val;
        for (Iterator i = values.iterator(); i.hasNext(); sum += val.get()) {
            val = (IntWritable) i.next();
        }
        this.result.set(sum);
        context.write(key, this.result);
    }
}
```

（4）编写 WordMain 驱动类。

要启动 MapReduce 作业，需要编写一个驱动类。编写完 WordMapper 类和 WordReducer 类后，接下来就要编写驱动类 WordMain。

示例 3

编写启动词频统计功能的驱动类。

关键代码：

```
public static void main(String[] args) throws IOException, ClassNotFoundException,
            InterruptedException {
    Configuration conf = new Configuration();
    String[] otherArgs = new String[]{"hdfs://hadoop:8020/wc/wordcount.txt","/outputwc"};
    if (otherArgs.length != 2) {
        System.exit(2);
    }
    Job job = Job.getInstance(conf, "WordCount");
    job.setJarByClass(WordCount.class);
    job.setMapperClass(WordCount.TokenizerMapper.class);
    job.setReducerClass(WordCount.IntSumReduce.class);
    job.setOutputKeyClass(Text.class);
    job.setOutputValueClass(IntWritable.class);
    FileInputFormat.addInputPath(job, new Path(otherArgs[0]));
    FileSystem fs = FileSystem.get(conf);
    Path path = new Path(otherArgs[1]);
    if (fs.isDirectory(path)) {
        fs.delete(path, true);
    }
    //设置文件的输出路径
    FileOutputFormat.setOutputPath(job, new Path(otherArgs[1]));
    System.exit(job.waitForCompletion(true) ? 0 : 1);
}
```

3. 提交 WordCount 作业到集群运行

MapReduce 作业代码编写完成后，需要打包提交到集群中运行。在前面已经介绍过如何搭建移动通信业务的 Hadoop 处理平台，读者可以在该平台上提交运行 MapReduce 作业。

示例 4

在 Hadoop 集群上提交运行词频统计作业并查看结果。关键步骤如下。

（1）使用 Maven 打包命令“mvn clean package -DskipTests”将前面创建的 mapreduceTest 工程打包成 hadoopwordcount.jar，并将 jar 文件上传到 Linux 服务器上的/home/hadoop/lib 目录下。

（2）准备测试数据。

```
[hadoop@hadoop tmp]# hdfs dfs -mkdir /wc
[hadoop@hadoop tmp]# hdfs dfs -put wordcount.txt /wc
```

（3）提交作业到集群运行。

```
[hadoop@hadoop lib]# hadoop jar /home/lib/hadoopwordcount.jar WordCount
```

（4）查看结果。

```
[hadoop@hadoop lib]hdfs dfs –text /outputwc/part-*
BigData    2
Hadoop     3
HDFS       1
Hello      3
MapReduce      1
Perfect    1
```

3.1.4 技能实训

使用 MapReduce API 完成用户手机流量统计功能。

需求：分析移动通信业务数据文件，统计每个手机号码的上行流量和下行流量。

输出结果字段：手机号码、上行总流量、下行总流量。

输入文件字段说明如表 3-2 所示。

表 3-2 输入文件字段说明

字段名	字段类型	字段描述	示例数据
reportTime	long	记录报告时间戳	1363157993044
msisdn	String	手机号码	18211585864
apmac	String	AP mac	94-71-AC-CD-E6-18:CMCC-EASY
acmac	String	AC mac	120.196.100.99
address	String	地址	bj
upPackNum	long	上行数据包个数	15
downPackNum	long	下行数据包个数	12
upPayLoad	long	上行总流量	1527
downPayLoad	long	下行总流量	2106
httpStatus	String	响应状态	200

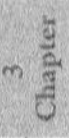

关键步骤：

（1）由于 map 和 reduce 输入/输出都为 key-value 键值对形式，所以必须将手机的流量信息（上行流量、下行流量）封装成一个 Bean 类，将这个类作为 value。

（2）对 Bean 类实现序列化和反序列化（实现 Writable 接口）。

（3）编写 MapReduce 任务，以 jar 包的形式执行。

任务 2　按号段统计手机号码

【任务描述】

本任务使用 MapReduce 实现按号段统计手机号码功能。

【关键步骤】

（1）理解 MapReduce 输入/输出格式。

（2）使用 Combiner 类优化 WordCount 程序。

（3）使用 Partitioner 操作实现按号段统计手机号码。

（4）理解 Shuffle 过程。

（5）自定义 RecordReader 类，分别统计移动通信数据文件中奇数行和偶数行的和。

3.2.1　MapReduce 输入/输出格式

使用 MapReduce 编程，只需定义好 map 和 reduce 函数的输入和输出<key,value>对的类型即可，不需关注如何输入文件块以及如何把键值对写入到 HDFS 文件块中，这部分工作其实是由 Hadoop 自带的输入和输出格式来处理的。Hadoop 根据输入文件格式的 RecordReader 来解析文件中的<key,value>对，默认情况下，一行代表一个<key,value>对。

1．**输入格式**

（1）InputFormat。InputFormat 是一个接口，定义了输入文件如何被 Hadoop 分块。定义如下。

```
public abstract class InputFormat<K,V>{
    public InputFormat(){
    }
    public abstract List<InputSplit> getSplit(JobContext context)
            throws IOException,InterruptedException;
    public abstract RecordReader<K,V> createRecordReader (InputSplit
             split,TaskAttemptContet context) thorws IOException,InterruptedException;
}
```

方法说明：

getSplit(JobContext context)方法负责将一个大数据在逻辑上拆分成一个或多个 InputSplit。每个 InputSplit 记录两个参数，第一个为这个分片数据的位置，第二个为这个分片数据的大小，很明显，InputSplit 并没有真正存储数据，只是提供了一个如何将数据分片的方法。

createRecordReader(InputSplit split,TaskAttemptContext context)方法根据 InputSplit 定义的方法，返回一个能够读取分片记录的 RecordReader。

（2）InputFormat 接口实现类。Hadoop 提供了很多 InputFormat 的实现类，其类继承结构如图 3.2 所示。

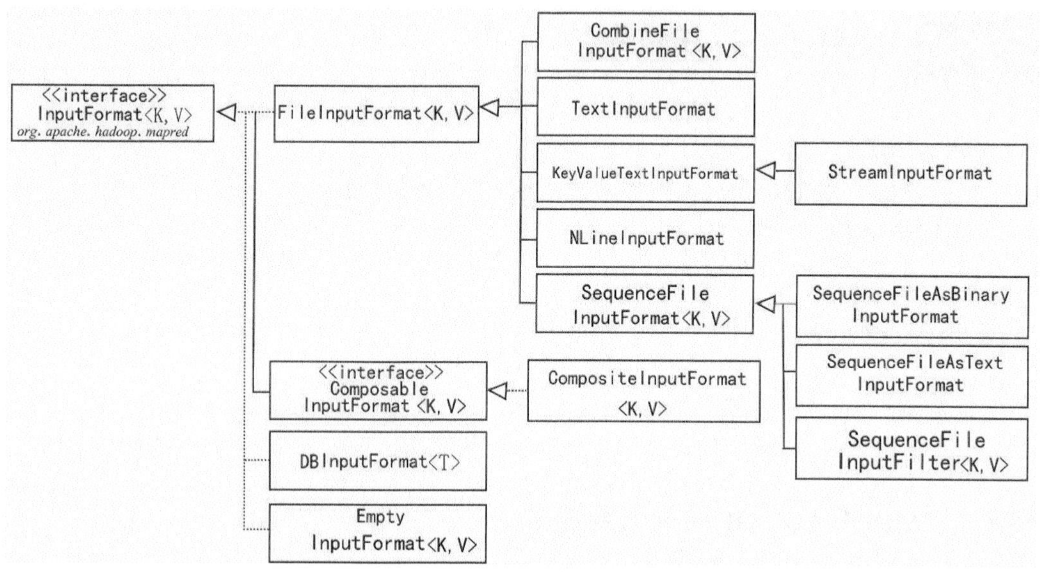

图3.2 InputFormat的类继承结构图

常用的 InputFormat 实现类如表 3-3 所示。

表 3-3 常用的 InputFormat 实现类

实现类	概述	key	value
TextInputFormat	Hadoop 默认的输入格式 读取文件行	行的字节偏移量（LongWritable）	行的内容（Text）
FileInputFormat	在 Hadoop 中，所有文件作为数据源的 InputFormat 实现的基类	用户自定义作业的输入路径	
KeyValueInputFormat	把行解析为<key,value>对，每一行均为一条记录	第一个分割符（缺省是“\t”）前的所有字符（Text）	第一个分隔符后剩下的内容（Text）

（3）设置 MapReduce 的输入格式。输入格式的设置可以在驱动类中使用 Job 对象的 setInputFormat()方法完成。当输入格式是 TextInputFormat 时，驱动类可以不设置输入格式。

2. **输出格式**

针对刚刚介绍的输入格式，Hadoop 都提供有对应的输出格式。输出格式用来确定如何将<key,value>对写入到 HDFS 文件块中。默认情况下，只有一个 Reduce，即输出只有一个文件，文件名为 part-r-00000。输出文件的个数与 Reduce 的个数一致。

（1）OutputFormat。OutputFormat 是一个接口，主要用于描述输出数据的格式，它能够将用户提供的<key,value>对写入到特定格式的文件中。OutputFormat 是 MapReduce

输出格式的基类，所有 MapReduce 输出都要继承 OutputFormat 抽象类。其类继承结构如图 3.3 所示。

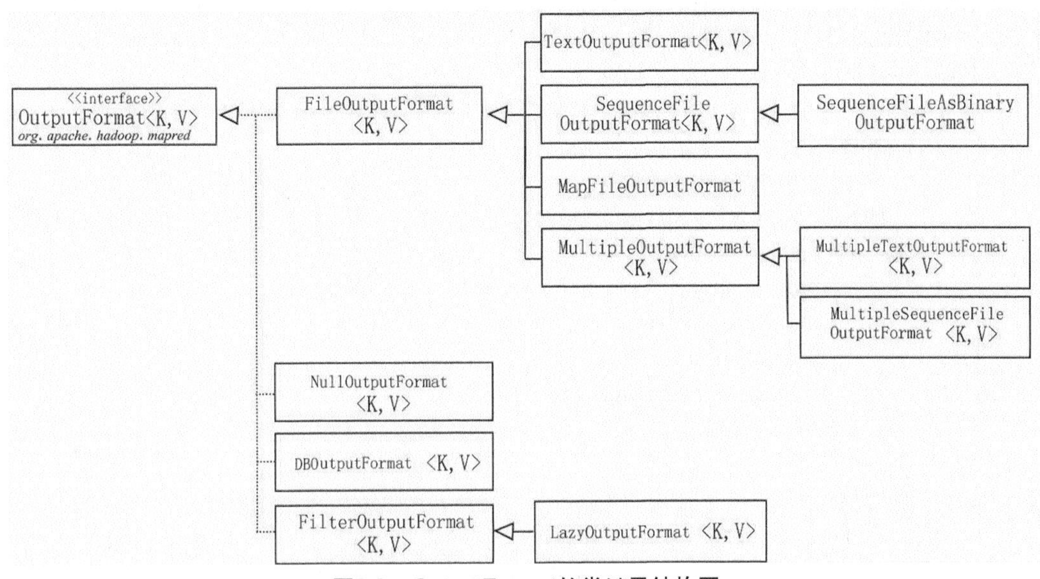

图3.3　OutputFormat的类继承结构图

（2）常用的 OutputFormat 实现类，如表 3-4 所示。

表 3-4　常用的 OutputFormat 实现类

实现类	概述
TextOutputFormat	Hadoop 默认的输出格式，将每条记录写成文本行，每个<key,value>对由制表符进行分隔
SequenceFileOutputFormat	输出二进制文件，如果输出需要作为后续 MapReduce 任务的输入，这是一种较好的输出格式

3.2.2　Combiner 类

1. Combiner 概述

Combiner 类是用来优化 MapReduce 的，它可以提高 MapReduce 的运行效率。在 MapReduce 作业运行过程中，通常每一个 Map 都会产生大量的本地输出，Combiner 的作用就是在 Map 端对输出结果先做一次合并，以减少传输到 Reduce 端的数据量。

Combiner 并没有自己的基类，而是继承 Reducer，它们对外的功能是一样的，只是使用的位置和使用的上下文不太一样，Combiner 操作发生在 Map 端。在某些情况下，Combiner 的加入不会影响原有的逻辑，即不影响最后的运行结果，影响的只是效率。

2. Combiner 在 WordCount 中的使用

示例 5

使用 Combiner 优化 WordCount 代码。

```
import org.apache.hadoop.*;
import java.io.IOException;
import java.util.StringTokenizer;
import java.net.URI;
public class MyWordCountCombinerApp {
    public static class TokenizerMapper extends Mapper<Object,Text,Text,IntWritable> {
        private final static IntWritable one = new IntWritable(1);
        private Text word = new Text();
        public void map(Object key, Text value, Mapper.Context context) throws IOException,
          InterruptedException {
            StringTokenizer itr = new StringTokenizer(value.toString());
            while (itr.hasMoreTokens()) {
                word.set(itr.nextToken());
                context.write(word, one);
            }
        }
    }
    public static class IntSumReducer extends Reducer<Text,IntWritable,Text,IntWritable>{
        private IntWritable result = new IntWritable();
        public void reduce(Text key,Iterable<IntWritable>values,Context context)
    throws IOException,InterruptedException{
             int sum = 0;
             for(IntWritable val:values){
                  sum+=val.get();
             }
             result.set(sum);
             context.write(key,result);
        }
    }
    public static void main(String[] args) throws Exception {
        Configuration conf = new Configuration();
        String[] otherArgs =
             new String[]{"hdfs://hadoop:8020/wordcount/input/word.txt","/wordcount/output"};
        if (otherArgs.length != 2) {
             System.err.println("Usage:Merge and duplicate removal <in> <out>");
             System.exit(2);
        }
        final FileSystem fileSystem = FileSystem.get(new URI(otherArgs[0]),conf);
        if(fileSystem.exists(new Path(otherArgs[1])))
        {
            fileSystem.delete(new Path(otherArgs[1]),true);
        }
        Job job = Job.getInstance(conf,"word count");
        job.setJarByClass(MyWordCountCombinerApp.class);
        job.setMapperClass(TokenizerMapper.class);
```

```
        //通过 job 设置 Combiner 处理类
        job.setCombinerClass(IntSumReducer.class);
        job.setReducerClass(IntSumReducer.class);
        job.setOutputKeyClass(Text.class);
        job.setOutputValueClass(IntWritable.class);
        FileInputFormat.addInputPath(job,new Path(otherArgs[0]));
        FileOutputFormat.setOutputPath(job,new Path(otherArgs[1]));
        System.exit(job.waitForCompletion(true) ? 0 : 1);
    }
}
```

3.2.3 Partitioner 类

上一节已经介绍了使用 Combiner 类可以提升 MapReduce 作业的执行效率，本节将介绍 MapReduce 的另一个优化组件——Partitioner 类。

Partitioner 的功能是在 Map 端对 key 进行分区。Map 端最终处理的<key,value>对需要发送到 Reduce 端去合并，合并的时候，相同分区的<key,value>对会被分配到同一个 Reduce 上，这个分配过程就是由 Partitioner（分区）决定的。

MapReduce 默认的 Partitioner 是 HashPartitioner。其计算方法如下。

① Partitioner 先计算 key 的散列值（通常是 MD5 值）。

② 通过 Reducer 个数执行取模运算：Key.hashCode%numReduce。

示例 6

按号段统计手机号码，手机号前三位相同的统计数据单独放在一个结果中。统计方式为按照 135 号段、136 号段、137 号段、138 号段、139 号段及其他号段分区，共 6 个分区。

关键代码：

（1）Mobile 实体类

```
import org.apache.hadoop.io.Writable;
import java.io.DataInput;
import java.io.DataOutput;
import java.io.IOException;
public class Mobile implements Writable {
    private String phoneNB;
    private long up_flow;
    private long down_flow;
    private long sum_flow;
    public Mobile() {
    }
    public Mobile(String phoneNB, long up_flow, long down_flow) {
        this.phoneNB = phoneNB;
        this.up_flow = up_flow;
        this.down_flow = down_flow;
```

```
        this.sum_flow = up_flow + down_flow;
    }
    public void write(DataOutput out) throws IOException {
        out.writeUTF(phoneNB);
        out.writeLong(up_flow);
        out.writeLong(down_flow);
        out.writeLong(sum_flow);
    }
    public void readFields(DataInput in) throws IOException {
        this.phoneNB = in.readUTF();
        this.up_flow = in.readLong();
        this.down_flow = in.readLong();
        this.sum_flow = in.readLong();
    }
    public String getPhoneNB() {
        return phoneNB;
    }
    public void setPhoneNB(String phoneNB) {
        this.phoneNB = phoneNB;
    }
    public long getUp_flow() {
        return up_flow;
    }
    public void setUp_flow(long up_flow) {
        this.up_flow = up_flow;
    }
    public long getDown_flow() {
        return down_flow;
    }
    public void setDown_flow(long down_flow) {
        this.down_flow = down_flow;
    }
    public long getSum_flow() {
        return sum_flow;
    }
    public void setSum_flow(long sum_flow) {
        this.sum_flow = sum_flow;
    }
    @Override
    public String toString() {
        return "" + up_flow + "\t" + down_flow + "\t" + sum_flow;
    }
```

```
}
```

（2）自定义分区

```
import org.apache.hadoop.mapreduce.Partitioner;
import org.apache.hadoop.io.Text;
import java.util.HashMap;
public class MyHashPartitioner extends Partitioner<Text, Mobile> {
    private static HashMap<String, Integer> areaMap = new HashMap<String, Integer>();
    static {
        areaMap.put("135", 0);
        areaMap.put("136", 1);
        areaMap.put("137", 2);
        areaMap.put("138", 3);
        areaMap.put("139", 4);
    }
    public int getPartition(Text text, Mobile value, int i) {
        Integer areCoder = areaMap.get(text.toString().substring(0, 3));
        if (areCoder == null) {
            areCoder = 5;
        }
        return areCoder;
    }
}
```

（3）MapReduce 功能

```
import java.io.IOException;
import org.apache.commons.lang.StringUtils;
import org.apache.hadoop.*;
public class MobileMapReduce {
    private final static String INPUT_PATH = "hdfs://hadoop:8020/mobile/input";
    private final static String OUTPUT_PATH = "hdfs://hadoop:8020/mobile/output";
    public static class MobileMapper   extends Mapper<LongWritable,Text,Text, Mobile>{
        public void map(LongWritable key, Text value,Mapper<LongWritable, Text, Text,
          Mobile>.Context context)throws IOException, InterruptedException {
            String line = value.toString();
            String[] fields = StringUtils.split(line, " ");
            String phoneNB = fields[1];
            long up_flow = Long.parseLong(fields[7]);
            long down_flow = Long.parseLong(fields[8]);
            // 封装数据为 kv 并输出
            context.write(new Text(phoneNB),
                        new Mobile(phoneNB, up_flow, down_flow));
        }
    }
    public static class FlowSumReducer extends Reducer<Text, Mobile, Text, Mobile> {
```

```
        protected void reduce(Text k2, Iterable<Mobile> v2s, Reducer<Text, Mobile, Text,
          Mobile>.Context context) throws IOException, InterruptedException {
            long up_flow = 0;
            long down_flow = 0;
            for (Mobile v2 : v2s) {
                up_flow += v2.getUp_flow();
                down_flow += v2.getDown_flow();
            }
            context.write(k2, new Mobile(k2.toString(), up_flow, down_flow));
        }
    }
    public static void main(String[] args) throws IOException,
      ClassNotFoundException, InterruptedException {
        Configuration conf = new Configuration();
        Job job = Job.getInstance(conf);
        job.setJarByClass(MobileMapReduce.class);
        job.setMapperClass(MobileMapper.class);
        job.setReducerClass(FlowSumReducer.class);
        job.setOutputKeyClass(Text.class);
        job.setOutputValueClass(Mobile.class);
        job.setPartitionerClass(MyHashPartitioner.class);
        job.setNumReduceTasks(6);
        FileInputFormat.setInputPaths(job, new Path(INPUT_PATH));
        FileSystem fs = FileSystem.get(conf);
        Path path = new Path(OUTPUT_PATH);
        if (fs.isDirectory(path)) {
            fs.delete(path, true);
        }
        FileOutputFormat.setOutputPath(job, new Path(OUTPUT_PATH));
        boolean res = job.waitForCompletion(true);
        System.exit(res?1:0);
    }
}
```

3.2.4 Shuffle 阶段

Shuffle 阶段是 MapReduce 奇迹发生的地方，它的大致处理过程就是把 Map 任务的输出结果有效地传送到 Reduce 端，也可以理解为 Shuffle 就是描述数据从 Map 任务输出到 Reduce 任务输入的这段过程。Shuffle 阶段分为 Map 端 Shuffle 和 Reduce 端 Shuffle 两个阶段，接下来分别介绍这两个阶段的功能和原理。

1. Map 端的 Shuffle

MapReduce 提交作业时，Map 端的 Shuffle 过程如下。

（1）拆分。InputSplit 将作业拆分成若干个 Map 任务。

（2）执行自定义 map()方法。Map 过程开始处理，Mapper 任务会接受输入分片，通过不断地调用 map()方法对数据进行处理。处理完毕后，转换为新的<key,value>对输出。

（3）缓存。Map 端的输出结果先写到内存中的一个缓冲区。

（4）分区（Partitioner）。在内存中进行分区，默认是 HashPartitioner，目的是将 Map 端的结果分给不同的 Reducer。

（5）排序。分区结束后，针对不同分区的数据，会按照 key 进行排序，要求 key 必须实现 WritableComparable 接口。

（6）分组。排序后的数据按 key 进行分组，将相同 key 的<key,value>对分到一个组中，最终，每个分组将会调用一次 reduce()方法。

（7）合并（Combiner）。排序分组后，相同 key 的数据组成一个列表，如果设置了 Combiner，就合并数据，可以减少写入磁盘的记录数。

（8）溢写（spill）。当内存中的数据达到一个特定的阈值时，系统会自动将记录溢写到磁盘中，形成一个 spill 文件。

（9）合并文件（Merge）。Map 过程结束时会把溢写出来的多个文件合并成一个。Merge 过程最多将 10 个（默认）文件同时合并成一个文件，多余的文件将分多次合并。

经过上面的流程，Map 端 Shuffle 过程完毕，数据都有序地存放在磁盘里，等待 Reducer 去拉取数据。

2. Reduce 端的 Shuffle

Reducer 在执行前，都是重复地拉取数据做合并，Reduce 端的 Shuffle 过程如下。

（1）复制过程。Reduce 端启动一些数据复制线程（Fetcher），通过 HTTP 方式请求获取 Map 任务的输出文件。

（2）合并阶段。这里的合并与 Map 端的合并类似，只是处理的数据是从不同 Map 端复制过来的。

（3）Reducer 的输入文件。经过不断地合并后，生成一个最终文件，当 Reducer 的输入文件已定，整个 Shuffle 过程才最终结束，然后就由 Reducer 执行，把结果放到 HDFS 上。

Shuffle
过程详解

3.2.5 自定义 RecordReader

1. RecordReader 概述

RecordReader 类用来加载数据并把数据转换为适合 Mapper 类里 map()方法读取的键值对，它决定了以怎样的方式从分片数据中读取一条记录，每读取一条记录都会调用 RecordReader 类。RecordReader 实例是由输入格式定义的，默认采用的输入格式是 TextInputFormat。通常使用的是 LineRecordReader 类，LineRecordReader 使用每行的偏移量作为 map 的 key，每行的内容作为 map 的 value。

2. 实现自定义 RecordReader 类

自定义 RecordReader 类的实现步骤如下：

（1）继承抽象类 RecordReader，主要是实现 nextKeyValue()方法。

（2）实现自定义 InputFormat 类，重写 InputFormat 类中的 createRecordReader()方法，返回自定义的 RecordReader 实例。

（3）在驱动类中配置 job.setInputFormatClass()，设置为自定义的 InputFormat 实例。

示例 7

自定义 RecordReader，实现统计移动通信数据文件中奇数行和偶数行的和。

关键代码：

```
import java.io.IOException;
import java.net.URI;
import java.net.URISyntaxException;
import java.util.List;
import org.apache.hadoop.*;
public class MyRecordReader {
    private final static String INPUT_PATH = "hdfs://hadoop:8020/inputsum";
    private final static String OUTPUT_PATH = "hdfs://hadoop:8020/outputsum";
    public static class DefRecordReader extends RecordReader<LongWritable, Text> {
        private long start;
        private long end;
        private long pos;
        private FSDataInputStream fin = null;
        private LongWritable key = null;
        private Text value = null;
        private LineReader reader = null;
        public void initialize(InputSplit split, TaskAttemptContext context) throws IOException,
            InterruptedException {
            FileSplit fileSplit = (FileSplit) split;
            start = fileSplit.getStart();
            end = start + fileSplit.getLength();
            Path path = fileSplit.getPath();
            Configuration conf = context.getConfiguration();
            FileSystem fs = path.getFileSystem(conf);
            fin = fs.open(path);
            fin.seek(start);
            reader = new LineReader(fin);
            pos = 1;
        }
        @Override
        public boolean nextKeyValue() throws IOException, InterruptedException {
            if (key == null) {
                key = new LongWritable();
            }
            key.set(pos);//设置 key
```

```
            if (value == null) {
                value = new Text();
            }
            if (reader.readLine(value) == 0) {
                return false;
            }
            pos++;
            return true;
        }
        @Override
        public LongWritable getCurrentKey() throws IOException, InterruptedException {
            return key;
        }
        @Override
        public Text getCurrentValue() throws IOException, InterruptedException {
            return value;
        }
        @Override
        public float getProgress() throws IOException, InterruptedException {
            return 0;
        }
        @Override
        public void close() throws IOException {
            fin.close();
        }
    }
    public static class MyFileInputFormat extends FileInputFormat<LongWritable, Text> {
        @Override
        public RecordReader<LongWritable, Text> createRecordReader(
            InputSplit split, TaskAttemptContext context) throws IOException,
             InterruptedException {
                return new DefRecordReader();
        }
        @Override
        protected boolean isSplitable(JobContext context, Path filename) {
            return false;
        }
    }
    public static class MyMapper extends Mapper<LongWritable, Text, LongWritable, Text> {
        @Override
        protected void map(LongWritable key, Text value, Context context) throws IOException,
            InterruptedException {
```

```
            context.write(key, value);
        }
    }
    public static class DefPartitioner extends Partitioner<LongWritable,Text>{
        @Override
        public int getPartition(LongWritable key, Text value, int numPartitions) {
            if(key.get()%2==0){
                key.set(1);
                return 1;
            }else {
                key.set(0);
                return 0;
            }
        }
    }
    public static class MyReducer extends Reducer<LongWritable, Text,Text, IntWritable>{
        Text write_key = new Text();
        IntWritable write_value = new IntWritable();
        @Override
        protected void reduce(LongWritable key, Iterable<Text> values, Context context) throws
                IOException, InterruptedException {
            int sum=0;
            for (Text val : values) {
                sum += Integer.parseInt(val.toString());
            }
            if(key.get()==0){
                write_key.set("奇数行之和");
            }else {
                write_key.set("偶数行之和");
            }
            write_value.set(sum);
            context.write(write_key, write_value);
        }
    }
    public static void main(String[] args) throws IOException, URISyntaxException,
                ClassNotFoundException, InterruptedException {
        Configuration conf = new Configuration();
        final FileSystem fileSystem = FileSystem.get(new URI(INPUT_PATH),conf);
        if(fileSystem.exists(new Path(OUTPUT_PATH))) {
            fileSystem.delete(new Path(OUTPUT_PATH),true);
        }
        Job job = Job.getInstance(conf, "Define RecordReader");
        job.setJarByClass(MyRecordReader.class);
```

```
            FileInputFormat.addInputPath(job, new Path(INPUT_PATH));
            job.setInputFormatClass(MyFileInputFormat.class);
            job.setMapperClass(MyMapper.class);
            job.setMapOutputKeyClass(LongWritable.class);
            job.setMapOutputValueClass(Text.class);
            job.setPartitionerClass(DefPartitioner.class);
            job.setReducerClass(MyReducer.class);
            job.setNumReduceTasks(2);
            job.setOutputKeyClass(Text.class);
            job.setOutputValueClass(IntWritable.class);
            FileOutputFormat.setOutputPath(job, new Path(OUTPUT_PATH));
            System.exit(job.waitForCompletion(true) ? 0 : 1);
        }
    }
```

3.2.6　技能实训

使用 MapReduce 编程找出微信共同好友。数据格式如下。

用户　好友 1　好友 2 …..
A B C D E F
B A C D E
C A B E
D A B E
……

结果输出：

A-B　D:E:C　表示 A 和 B 的共同好友为 D、E 和 C
A-C　B:E　表示 A 和 C 的共同好友是 B 和 E
……

关键步骤：

（1）求出每一个人都是哪些人的共同好友。

（2）把这些人（拥有共同好友的人）作为 key，其好友作为 value 输出。

任务 3　使用 MapReduce 编写应用案例

【任务描述】

本任务使用 MapReduce 实现文件 join 操作、排序功能以及二次排序功能。

【关键步骤】

（1）使用 MapReduce 实现类似 SQL 中的 join 操作。

（2）使用 MapReduce 实现排序功能。

（3）使用 MapReduce 实现二次排序功能。

3.3.1 使用 MapReduce 实现 join 操作

1. 需求

在关系数据库中，join 是非常常见的操作。在海量数据处理过程中，也不可避免地会碰到这种类型的处理需求。例如，在数据分析时需要连接从不同数据源中获取的数据。不同于传统的单机模式，在分布式环境下采用的 MapReduce 编程模型，也有相应的处理措施和优化方法。本节通过两个示例分别演示 Map 端 join 和 Reduce 端 join 的方法。

2. Map 端 join

示例 8

使用 MapReduce 编程模型实现学生信息和成绩信息的关联查询。

学生信息文件 students.txt 的数据如下。

```
2016001,Join
2016002,Abigail
2016003,Abby
2016004,Alexandra
2016005,Cathy
2016006,Katherine
```

成绩信息文件 student_score.txt 的数据如下。

```
2016001,YY,60
2016001,SX,88
2016001,YW,91
2016002,SX,77
2016002,YW,33
```

通过 MapReduce 输出的结果如下。

```
2016001,Join,YY,60
2016001,Join,SX,88
2016001,Join,YW,91
2016002,Abigail,SX,77
2016002,Abigail,YW,33
```

关键代码：

```
import java.io.BufferedReader;
import java.io.IOException;
import java.io.InputStreamReader;
import java.net.URI;
import java.util.*;
import org.apache.commons.lang.StringUtils;
import org.apache.hadoop.*;
/**
* MapReduce 任务实战 Map Join
*/
```

```
public class MapJoinDirectOutPutJob extends Configured implements Tool {
    private static String STUDENTNAME_PARAM ="STUDENTNAME_PARAM";
    private static String INPATH_SCORE = "joinjob/student_score.txt";
    private static String OUTPATH = "joinjob/output";
    public static class MapJoinDirectOutPutMapper extends
                    Mapper<LongWritable,Text,NullWritable,Text>{
        private BufferedReader br = null;
        private Map <String,String> map =new HashMap<String,String>();
        private Text newValue = new Text();
        @Override
        protected void map(LongWritable key, Text value, Context context)
          throws IOException, InterruptedException {
            String [] words = StringUtils.split(value.toString(),',');
            String name = map.get(words[0]);
            newValue.set(words[0]+","+name +","+words[1]+","+words[2]);
            context.write(NullWritable.get(), newValue);
        }
        @Override
        protected void setup(Context context) throws IOException,
          InterruptedException {
            Configuration conf = context.getConfiguration();
            FileSystem fs = FileSystem.getLocal(conf);
             br = new BufferedReader(new InputStreamReader
                                        (fs.open(new Path("studentLink.txt"))));
             String current;
             while((current= br.readLine())!=null){
                    String [] words = current.split(",");
                    map.put(words[0],words[1]);
             }
             br.close();
        }
    }
    @Override
    public int run(String[] args) throws Exception {
        Job job = Job.getInstance(getConf(),"MapJoinDirectOutPutJob");
        job.setJarByClass(getClass());
        Configuration conf = job.getConfiguration();
        conf.set(STUDENTNAME_PARAM, args[0]);
        Path in = new Path(INPATH_SCORE);
        Path out = new Path(OUTPATH);
        FileSystem.get(conf).delete(out,true);
        FileInputFormat.setInputPaths(job, in);
        FileOutputFormat.setOutputPath(job, out);
```

```
        job.setInputFormatClass(TextInputFormat.class);
        job.setOutputFormatClass(TextOutputFormat.class);
        job.setMapperClass(MapJoinDirectOutPutMapper.class);
        job.setMapOutputKeyClass(NullWritable.class);
        job.setMapOutputValueClass(Text.class);
        job.setNumReduceTasks(0);
        URI uri = new URI("hdfs://hadoop:8020/user/train/joinjob/students.txt#studentLink.txt");
        job.addCacheFile(uri);
        return job.waitForCompletion(true) ?0:1;
    }
    public static void main(String [] args){
        int r = 0;
        try{
            r = ToolRunner.run(new Configuration(),new MapJoinDirectOutPutJob(),args);
        }catch(Exception e){
            e.printStackTrace();
        }
        System.exit(r);
    }
}
```

3. Reduce 端 join

示例 9

使用 MapReduce 编程模型实现员工信息和部门信息的关联查询。

员工信息文件 EMP.txt 的数据如下。

```
zhang   male    20   1
li      female  25   2
wang    female  30   3
zhou    male    35   2
```

部门信息文件 DEP.txt 的数据如下。

```
1       Sales
2       Dev
3       Mgt3
```

通过 MapReduce 输出的结果如下。

```
EmpJoinDep [Name=zhang, Sex=male, Age=20, DepName=Sales]
EmpJoinDep [Name=zhou, Sex=male, Age=35, DepName=Dev]
EmpJoinDep [Name=li, Sex=female, Age=25, DepName=Dev]
EmpJoinDep [Name=wang, Sex=female, Age=30, DepName=Mgt]
```

关键代码：

（1）实体类定义

```
import java.io.DataInput;
import java.io.DataOutput;
import java.io.IOException;
```

```
import org.apache.hadoop.io.WritableComparable;
public class EmpJoinDep implements WritableComparable {
    private String Name = "";
    private String Sex = "";
    private int Age = 0;
    private int DepNo = 0;
    private String DepName = "";
    private String table = "";
    public EmpJoinDep() {
    }
    public EmpJoinDep(EmpJoinDep empJoinDep) {
        this.Name = empJoinDep.getName();
        this.Sex = empJoinDep.getSex();
        this.Age = empJoinDep.getAge();
        this.DepNo = empJoinDep.getDepNo();
        this.DepName = empJoinDep.getDepName();
        this.table = empJoinDep.getTable();
    }
    public String getName() {
        return Name;
    }
    public void setName(String name) {
        Name = name;
    }
    public String getSex() {
        return Sex;
    }
    public void setSex(String sex) {
        this.Sex = sex;
    }
    public int getAge() {
        return Age;
    }
    public void setAge(int age) {
        this.Age = age;
    }
    public int getDepNo() {
        return DepNo;
    }
    public void setDepNo(int depNo) {
        DepNo = depNo;
    }
    public String getDepName() {
```

```
        return DepName;
    }
    public void setDepName(String depName) {
        DepName = depName;
    }
    public String getTable() {
        return table;
    }
    public void setTable(String table) {
        this.table = table;
    }
    @Override
    public void write(DataOutput out) throws IOException {
        out.writeUTF(Name);
        out.writeUTF(Sex);
        out.writeInt(Age);
        out.writeInt(DepNo);
        out.writeUTF(DepName);
        out.writeUTF(table);
    }
    @Override
    public void readFields(DataInput in) throws IOException {
        this.Name = in.readUTF();
        this.Sex = in.readUTF();
        this.Age = in.readInt();
        this.DepNo = in.readInt();
        this.DepName = in.readUTF();
        this.table = in.readUTF();
    }
    @Override
    public int compareTo(Object o) {
        return 0;
    }
    @Override
    public String toString() {
        return "EmpJoinDep [Name=" + Name + ", Sex=" + Sex + ", Age=" + Age
                + ", DepName=" + DepName + "]";
    }
}
```

（2）MapReduce 功能实现

```
import java.io.IOException;
import java.net.URI;
import java.util.*;
```

```
import org.apache.hadoop.*;
public class ReduceJoin {
    private final static String INPUT_PATH = "hdfs://hadoop:8020/inputjoin";
    private final static String OUTPUT_PATH = "hdfs://hadoop:8020/outputmapjoin";
    public static class MyMapper extends Mapper<LongWritable, Text, IntWritable, EmpJoinDep>{
        private EmpJoinDep empJoinDep = new EmpJoinDep();
        @Override
        protected void map(LongWritable key, Text value, Context context)
                throws IOException, InterruptedException {
            String[] values = value.toString().split("\\s+");
            if(values.length==4){
                empJoinDep.setName(values[0]);
                empJoinDep.setSex(values[1]);
                empJoinDep.setAge(Integer.parseInt(values[2]));
                empJoinDep.setDepNo(Integer.parseInt(values[3]));
                empJoinDep.setTable("EMP");
                context.write(new IntWritable(Integer.parseInt(values[3])), empJoinDep);
            }
            if(values.length==2){
                empJoinDep.setDepNo(Integer.parseInt(values[0]));
                empJoinDep.setDepName(values[1]);
                empJoinDep.setTable("DEP");
                context.write(new IntWritable(Integer.parseInt(values[0])), empJoinDep);
            }
        }
    }
    public static class MyReducer extends Reducer<IntWritable, EmpJoinDep, NullWritable,
      EmpJoinDep>{
        @Override
        protected void reduce(IntWritable key, Iterable<EmpJoinDep> values,
                                    Context context)
                throws IOException, InterruptedException {
            String depName = "";
            List<EmpJoinDep> list = new LinkedList<EmpJoinDep>();
            for (EmpJoinDep val : values) {
                list.add(new EmpJoinDep(val));
                if(val.getTable().equals("DEP")){
                    depName = val.getDepName();
                }
            }
            for (EmpJoinDep listjoin : list) {
                if(listjoin.getTable().equals("EMP")){
                    listjoin.setDepName(depName);
```

```
                    context.write(NullWritable.get(), listjoin);
                }
            }
        }
    }
    public static void main(String[] args) throws Exception {
        Configuration conf = new Configuration();
        final FileSystem fileSystem = FileSystem.get(new URI(INPUT_PATH),conf);
        if(fileSystem.exists(new Path(OUTPUT_PATH))){
            fileSystem.delete(new Path(OUTPUT_PATH),true);
        }
        Job job = Job.getInstance(conf, "Reduce Join");
        job.setJarByClass(ReduceJoin.class);
        FileInputFormat.addInputPath(job, new Path(INPUT_PATH));
        job.setMapperClass(MyMapper.class);
        job.setMapOutputKeyClass(IntWritable.class);
        job.setMapOutputValueClass(EmpJoinDep.class);
        job.setReducerClass(MyReducer.class);
        job.setOutputKeyClass(NullWritable.class);
        job.setOutputValueClass(EmpJoinDep.class);
        FileOutputFormat.setOutputPath(job, new Path(OUTPUT_PATH));
        System.exit(job.waitForCompletion(true) ? 0 : 1);
    }
}
```

3.3.2 使用 MapReduce 实现排序功能

1. 需求

要求对用户使用的流量进行排序，文件中每一行内容均为用户使用的流量，输出结果包含两个间隔的数字，其中，第一个数字代表用户流量数据在原始数据集中的排位，第二个数字为用户流量数据。用户流量数据格式和输出结果如表 3-5 所示。

表 3-5 用户流量数据格式和输出结果表

用户流量	输出结果
500	1 490
490	2 500
510	3 510
636	4 636

2. 实现原理

在 MapReduce 中，默认可以对 IntWritable 类型和 Text 类型的 key 进行排序。本案例使用 IntWritable 类型数据结构，在 Map 端将输入的数据转换成 IntWritable 类型，作为 key 值输出，Reduce 端拿到<key,value-list>之后，将输入的 key 作为 value 输出，Reduce 端的 key 为全局变量，用于记录 key 的当前位次。

3. 代码实现

示例 10

对用户使用的数据流量进行排序。

关键代码：

```
import java.io.IOException;
import org.apache.hadoop.conf.Configuration;
import org.apache.hadoop.fs.Path;
import org.apache.hadoop.io.*;
import org.apache.hadoop.mapreduce.*;
import org.apache.hadoop.fs.FileSystem;
/**
* 使用 MapReduce 实现排序
* Input 路径下可以存放多个数字内容文件，会合并排序
*/
public class SortApp {
    //map 将输入的 value 转化成 IntWritable 类型，作为输出的 key
    public static class SortMapper extends Mapper<LongWritable, Text, IntWritable, IntWritable> {
        private   IntWritable data = new IntWritable();
        private   final IntWritable one = new IntWritable(1);
        @Override
        protected void map(LongWritable key, Text value, Context context)
          throws IOException, InterruptedException {
            String line = value.toString();
            data.set(Integer.parseInt(line));
            context.write(data, one);
        }
    }
    public static class SortReducer extends Reducer<IntWritable, IntWritable, IntWritable,
      IntWritable> {
        private IntWritable linenumber = new IntWritable(1);
        @Override
        protected void reduce(IntWritable key, Iterable<IntWritable> values,
                              Context context) throws IOException, InterruptedException {
            for (IntWritable value : values) {
                    context.write(linenumber, key);
                    linenumber.set(linenumber.get() + 1);
            }
        }
    }
    public static void main(String[] args) throws IllegalArgumentException,
      IOException, ClassNotFoundException, InterruptedException {
```

```
        Configuration conf = new Configuration();
        Job job = Job.getInstance(conf);
        job.setJarByClass(SortApp.class);
        job.setMapperClass(SortMapper.class);
        job.setReducerClass(SortReducer.class);
        job.setOutputKeyClass(IntWritable.class);
        job.setOutputValueClass(IntWritable.class);
        FileSystem fs = FileSystem.get(conf);
        Path path = new Path("/sort/output");
        if(fs.isDirectory(path)){
            fs.delete(path, true);
        }
        FileInputFormat.setInputPaths(job, new Path("/sort/input"));
        FileOutputFormat.setOutputPath(job, new Path("/sort/output"));
        job.waitForCompletion(true);
    }
}
```

4. 提交运行并查看结果

（1）使用 Maven 进行打包，并上传到/home/hadoop/lib 目录下。打包命令如下。

```
mvn clean package -DskipTests
```

（2）将测试数据上传到 HDFS。命令格式如下。

```
[hadoop@hadoop tmp]# hdfs dfs -mkdir -p /sort/input
[hadoop@hadoop tmp]# hdfs dfs -put sort.txt /sort/input
```

（3）提交 MapReduce 作业到集群运行。命令如下。

```
[hadoop@hadoop tmp]# hadoop jar /home/hadoop/lib/SortApp.jar SortApp
```

（4）查看运行结果。

```
[hadoop@hadoop tmp]# hdfs dfs -cat /sort/output/part-r-00000
1       490
2       500
3       510
4       636
```

3.3.3 使用 MapReduce 实现二次排序功能

1. 需求

在 MapReduce 操作时，Shuffle 阶段会多次根据 key 值排序，但是在 Shuffle 分组后，相同 key 值的 value 序列的顺序是不确定的，如果对 value 值也进行排序，这就需要用到二次排序。

对移动通信数据文件中的流量数据（每行两列，第一列为上行流量，第二列为总流量，列与列之间的分隔符是制表符）进行二次排序，输出结果先按照第一列的升序排列，如果第一列的值相等，就按照第二列的升序排列。数据格式如表 3-6 所示。

表 3-6　手机号的流量数据格式和输出结果表

输入数据	输出结果
90　180	60　98
60　395	60　395
60　98	78　110
78　110	90　180

2. 实现原理

二次排序可以分为以下几个阶段。

（1）Map 起始阶段。使用 job.setInputFormatClass 定义的 InputFormat 将输入的数据集分割成小数据块 split，同时 InputFormat 提供一个 RecordReader 的实现。

（2）Map 最后阶段。先调用 job.setPartitionerClass()对 Mapper 的输出结果进行分区，每个分区映射到一个 Reducer，每个分区内又调用 job.setSortComparatorClass()设置 key 比较函数类进行排序，可以看到，这本身就是一个二次排序。如果没有通过 job.setSortComparatorClass()设置 key 比较函数类，则使用 key 实现的 compareTo()方法。

（3）Reduce 阶段。Reduce()方法接受所有映射到这个 Reduce 的 map 输出后，也会调用 job.setSortComparatorClass()方法设置 key 比较函数类，对所有数据进行排序。然后构造每一个 key 值对应的 value 迭代器。这时就要用到分组，使用 job.setGroupingComparatorClass()方法设置分组函数类。只要比较的两个 key 值相同，它们就属于同一组，将它们的 value 值放在一个 value 迭代器中，而这个迭代器的 key 值使用属于同一个组的所有 key 值中的第一个。最后就是进入 Reducer 的 reduce()方法，该方法的输入是所有 key 值和它对应的 value 迭代器，同样注意输入和输出的类型必须与自定义的 Reduce 中声明的一致。

3. 代码实现

示例 11

对用户使用流量进行二次排序。

关键代码：

（1）自定义 IntPair 类实现 WritableComparable 接口，实现当 key 值相同时对 value 进行排序。

```
import java.io.DataInput;
import java.io.DataOutput;
import java.io.IOException;
import org.apache.hadoop.io.WritableComparable;
public class IntPair implements WritableComparable<IntPair> {
    private int first=0;
    private int second = 0;
    public void set(int left,int right){
        first = left;
        second = right;
    }
```

```
        public int getFirst(){
                return first;
        }
        public int getSecond(){
                return second;
        }
        @Override
        public void readFields(DataInput in) throws IOException {
                first = in.readInt();
                second = in.readInt();
        }
        @Override
        public void write(DataOutput out) throws IOException {
                out.writeInt(first);
                out.writeInt(second);
        }
        @Override
        public int hashCode() {
                return first+"".hashCode()+second+"".hashCode();
        }
        @Override
        public boolean equals(Object right){
                if(right instanceof IntPair){
                        IntPair r = (IntPair)right;
                        return r.first==first&&r.second==second;
                }else{
                        return false;
                }
        }
        @Override
        public int compareTo(IntPair o) {
                if(first!=o.first){
                        return first - o.first;
                }else if(second!=o.second){
                        return second - o.second;
                }else{
                        return 0;
                }
        }
}
```

（2）完成二次排序功能。

```
import java.io.IOException;
import java.net.URI;
import java.net.URISyntaxException;
import java.util.StringTokenizer;
```

```
import org.apache.hadoop.*;
public class SecondarySortApp {
    public static class MyMapper extends Mapper<LongWritable, Text, IntPair,IntWritable>{
        private final IntPair key = new IntPair();
        private final IntWritable value = new IntWritable();
        @Override
        protected void map(LongWritable inKey, Text inValue,Context context)
          throws IOException, InterruptedException {
            StringTokenizer itr = new StringTokenizer(inValue.toString());
            int left = 0;
            int right = 0;
            if(itr.hasMoreTokens()){
                left = Integer.parseInt(itr.nextToken());
                if(itr.hasMoreTokens()){
                    right = Integer.parseInt(itr.nextToken());
                }
                key.set(left, right);
                value.set(right);
                context.write(key, value);
            }
        }
    }
    public static class GroupingComparator implements RawComparator<IntPair>{
        public int compare(IntPair o1, IntPair o2) {
            Int first1 = o1.getFirst();
            int first2 = o2.getFirst();
            return first1 - first2;
        }
        public int compare(byte[] b1, int s1, int l1, byte[] b2, int s2, int l2) {
            return WritableComparator.compareBytes(b1, s1, Integer.SIZE/8, b2, s2, Integer.SIZE/8);
        }
    }
    public static class MyReducer extends Reducer<IntPair,IntWritable,Text,IntWritable>{
        private static final Text SEPARATOR = new Text("--------");
        private final Text first = new Text();
        @Override
        protected void reduce(IntPair key, Iterable<IntWritable> values,Context context)
          throws IOException, InterruptedException {
            context.write(SEPARATOR, null);
            first.set(Integer.toString(key.getFirst()));
            for(IntWritable value:values){
                context.write(first, value);
            }
        }
    }
```

```
    public static void main(String[] args) throws IOException, URISyntaxException,
      ClassNotFoundException, InterruptedException {
        String INPUT_PATH="hdfs://hadoop000:8020/secondsort";
        String OUTPUT_PATH="hdfs://hadoop000:8020/outputsecondsort";
        Configuration conf = new Configuration();
        final FileSystem filesystem = FileSystem.get(new URI(INPUT_PATH),conf);
        if(filesystem.exists(new Path(OUTPUT_PATH))){
                filesystem.delete(new Path(OUTPUT_PATH),true);
        }
        Job job = Job.getInstance(conf, "SecondarySortAPP");
        job.setJarByClass(SecondarySortApp.class);
        FileInputFormat.setInputPaths(job, new Path(INPUT_PATH));
        FileOutputFormat.setOutputPath(job, new Path(OUTPUT_PATH));
        job.setMapperClass(MyMapper.class);
        job.setReducerClass(MyReducer.class);
        job.setGroupingComparatorClass(GroupingComparator.class);
        job.setMapOutputKeyClass(IntPair.class);
        job.setMapOutputValueClass(IntWritable.class);
        job.setOutputKeyClass(Text.class);
        job.setOutputValueClass(IntWritable.class);
        job.setInputFormatClass(TextInputFormat.class);
        job.setOutputFormatClass(TextOutputFormat.class);
        System.exit(job.waitForCompletion(true)?0:1);
    }
}
```

4. 提交运行并查看结果

（1）使用 Maven 进行打包，并上传到/home/hadoop/lib 目录下。打包命令如下。

```
mvn clean package -DskipTests
```

（2）将测试数据上传到 HDFS。命令格式如下。

```
[hadoop@hadoop tmp]# hdfs dfs -put secondsort.txt /secondsort
```

（3）提交 MapReduce 作业到集群运行。命令如下。

```
[hadoop@hadoop tmp]# hadoop jar /home/hadoop/lib/SecondarySortApp.jar SecondarySortApp
```

（4）查看运行结果。

```
[hadoop@hadoop tmp]# hdfs dfs -cat /outputsecondsort/part-r-00000
--------
60    98
60    395
--------
78    110
--------
90    180
```

更多 MapReduce 操作案例请扫描二维码。

MapReduce 应用案例

3.3.4 技能实训

使用 MapReduce 编程实现自定义计数器。假设有 log.txt 文件，内容如下。

响应时间	IP	请求类型	响应状态	
23	192.168.10.135	GET	200	
26	192.168.10.124		404	不合法
29	192.168.10.127	GET	200	
31	192.168.10.129	GET	505	
10	192.168.10.123		404	不合法

需求：4 列都有内容的一行日志将认为是标准日志，即合法日志。请编写计数器，统计文件中不合法日志的行数。

关键步骤：

（1）对于少于 4 个字段的日志自定义一个名为 ERROR_LOG_TIME 的计数器来统计。

（2）使用 increment(1)方法来统计个数，出现一次，ERROR_LOG_TIME 计数器就加 1。

本章总结

➢ MapReduce 是 Hadoop 架构中的分布式计算框架，可以使用 MapReduce API 方便地开发分布式程序。

➢ 简单的 MapReduce 开发，用户只需要编写 Mapper 类、Reducer 类及驱动类即可。

➢ MapReduce 提供了多种输入格式（InputFormat）和输出格式（OutputFormat）。

➢ MapReduce API 中提供了 Combiner 和 Partitioner 组件，用于优化 MapReduce 执行效率。

本章作业

一、简答题

1. 简述 MapReduce 执行过程。

2. 简述 MapReduce 中 Combiner 和 Partitioner 的使用。

二、编码题

1. 使用 MapReduce 编程实现 SQL 中的去重（Distinct）操作。数据内容如下。

```
hello    world
hello    hadoop
hello    mapreduce
hello    world
hello    hadoop
hello world
```

输出的结果如下。

```
hello    hadoop
hello    mapreduce
```

```
hello	world
```

2．使用 MapReduce 编程实现统计网页的 PV（页面浏览量）。数据内容如下。

```
192.168.1.1
192.168.1.2
192.168.1.3
192.168.1.2
…
```

3．使用 MapReduce 编程实现统计网页的 UV（独立访客访问数）。数据内容如上题所示。

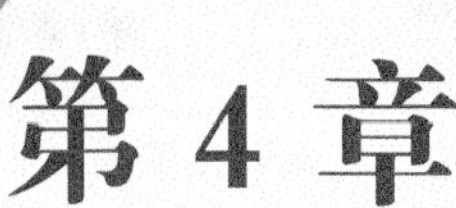

Hadoop YARN

技能目标

- 了解 YARN 产生背景
- 理解 YARN 架构设计
- 理解 YARN 的高可用架构设计
- 掌握使用 YARN 提交 MapReduce 作业

本章任务

任务 1　在 YARN 集群上运行 MapReduce 作业
任务 2　配置 YARN 容错

本章资源下载

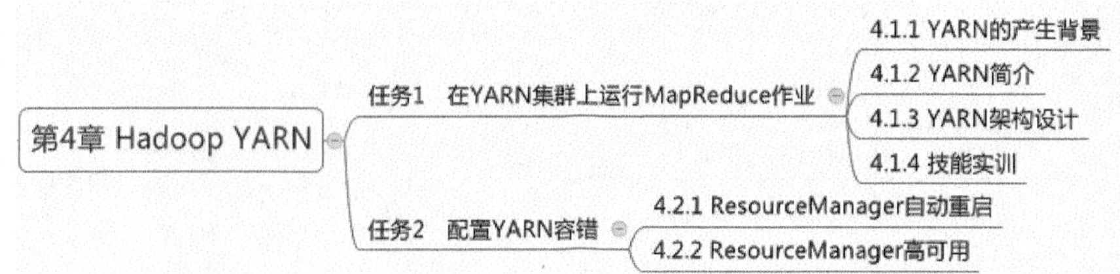

Apache Hadoop YARN（Yet Another Resource Negotiator）是 Hadoop2.0 中新引入的资源管理器，它为集群在资源利用率和统一管理以及数据共享方面带有极大的好处。YARN 是一个通用的资源管理系统，也就是说，在 YARN 上不仅可以提交 MapReduce 作业，而且可以提交 Spark 等作业。本章主要介绍 YARN 的产生背景、架构设计以及容错机制。

任务 1 在 YARN 集群上运行 MapReduce 作业

【任务描述】

本任务主要了解 YARN 的产生背景、架构设计，掌握在 YARN 上提交 MapReduce 作业。

【关键步骤】

（1）了解 YARN 的产生背景。

（2）理解 YANR 的架构及各组件的功能职责。

（3）掌握在 YARN 上提交 MapReduce 作业。

4.1.1 YARN 的产生背景

1．MapReduce1.0

在 Hadoop1.x 中，MapReduce 1.0 的架构图如图 4.1 所示。

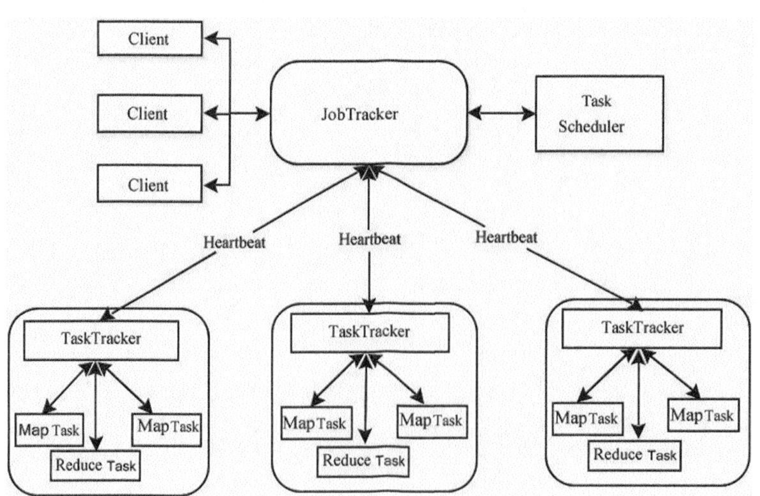

图4.1 MapReduce1.0的架构图

从图 4.1 所示的架构图中可以看出，在 Hadoop1.x 中，MapReduce 采用的是 Master/Slave 架构，在集群中表现为一个 JobTracker 和多个 TaskTracker。具体功能职责如下。

JobTracker 负责资源管理和作业调度，TaskTracker 则定期向 JonTracker 发送心跳信息，汇报本节点的健康状况、资源使用情况，以及任务的执行情况，同时还可以接收并执行 JobTracker 的命令，例如启动任务或者杀死任务等。

随着集群规模的发展，这样的架构会存在以下问题。

（1）JobTracker 单点故障。整个集群中只有一个 JobTracker，会存在单点故障的隐患，当 JobTracker 出现故障后，代表着整个集群将不能提供服务。

（2）JobTracker 节点压力大。从图 4.1 中可以看出，JobTracker 不仅要处理来自客户端的请求，还要接收大量 TaskTracker 节点的请求。JobTracker 承载职责过多，基本上整个集群的事情都由 JobTracker 来管理。

（3）不易于扩展。由于 JobTracker 是单节点，使它成为系统的最大瓶颈，严重制约了 Hadoop 集群的扩展性。

（4）只支持 MapReduce 作业。在 Hadoop1.x 中，只支持提交 MapReduce 作业，不支持提交其他框架的作业。随着互联网的高速发展，出现了一些新的计算框架，包括内存计算框架、流式计算框架和迭代式计算框架等，而 MapReduce1.0 并不支持多种计算框架并存。

2. 资源利用率

在没有 YARN 之前，由于 Hadoop1.x 不支持其他框架的作业，导致需要根据不同的框架去搭建多个集群，这样的方式造成了集群管理复杂，需要多个管理员来管理，无形中增加了运维成本；同时造成集群资源利用率很低。在实际应用中，通常 Hadoop 集群比较忙的时候，Spark 集群就比较闲；Spark 集群比较忙的时候，Hadoop 集群就比较闲。采用一个计算框架一个集群的方式，无法高效地利用集群资源。

3. 资源共享

数据共享是优化大数据集群的非常重要的方式。随着信息时代下数据的爆炸式增长，如果仍然采用一个集群一个计算框架的模式，跨集群的数据移动将会很频繁，不仅花费时间更长，硬件成本也会大大增加。所以就需要实现集群中的数据共享，也就是共享集群模式。该模式可以让多种框架共享数据和硬件资源，减少数据移动带来的开销，并且该模式只需要少数管理员即可完成多个框架的统一管理，减少了运维成本。

正是由于 Hadoop1.x 中存在上述问题，才使得 YARN 得以诞生。YARN 允许不同的计算框架运行在同一个集群上。

4.1.2 YARN 简介

1. YARN 概述

YARN 是随着 Hadoop 的发展而催生的新框架，目前已经成为 Hadoop2.x 架构中的核心构成。它的出现取代了 Hadoop1.x 中的 JobTracker，解决了 JobTracker 存在的单点故障、任务过重等问题。最主要的是在 YARN 集群上，不仅能运行 MapReduce 作业，而且能运行 Spark、Storm 等作业，YARN 是一个通用的资源管理系统。

2. YARN 架构

YARN 的架构也是典型的 Master/Slave 架构，由 Clinet、ResourceManager（简称 RM）、NodeManager（简称 NM）组成。YARN 的基本思想是将 JobTracker 的两个主要职责——资源管理和作业调度管理分别交给两个角色负责，一个是全局的 ResourceManager，一个是表示每个应用的 ApplicationMaster（简称 AM）。一个集群中，通常包含一个 ResourceManager 和多个 NodeManager。ResourceManager 以及每个节点一个的 NodeManager 组成新的资源管理系统。YARN 的架构如图 4.2 所示。

YARN 架构

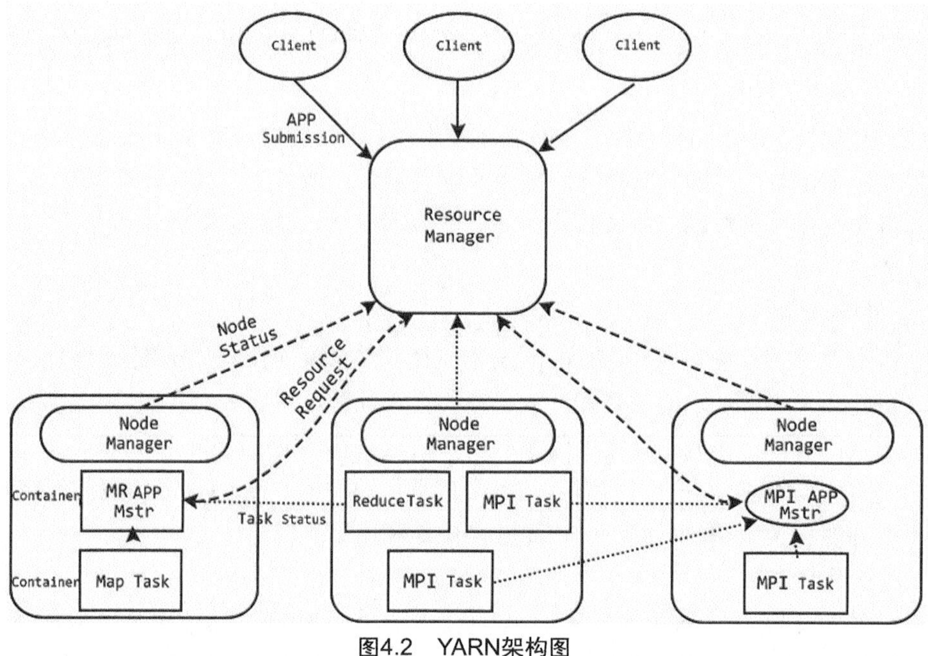

图4.2 YARN架构图

（1）ResourceManager（RM）。RM 是一个全局的资源管理器，整个集群中只有一个，负责整个系统的资源管理和作业调度。它主要由两个组件构成：调度器（Scheduler）和应用程序管理器（Applications Manager，ASM）。

➢ 调度器根据容量、队列等限制条件（如每个队列分配一定的资源，最多执行一定数量的作业等），将系统中的资源分配给正在运行的应用程序。调度器不做任何与应用程序相关的工作，仅根据各个应用程序的资源需求进行资源分配。另外，调度器是一个可插拔的组件，用户可以根据实际需要设计新的调度器。YARN 提供了多种直接可用的调度器，比如 Fair Schedule 和 Capacity Schedule 等。

➢ 应用程序管理器负责管理整个系统中的所有应用程序，包括应用程序的提交、与调度器协商资源以启动 ApplicationMaster、监控 ApplicationMaster 运行状态并在失败时启动它等。

（2）ApplicationMaster（AM）。用户提交的每一个应用程序都包含一个 AM，它主要负责应用程序的管理，具体功能如下。

➢ 进行数据切分。

➢ 为应用程序向 RM 调度器申请资源（以 Container 表示），并分配给内部任务。

➢ 与 NM 通信，启动/停止任务。

➢ 监控任务运行状态，并提供容错机制（任务失败时重新为任务申请资源，然后重启任务）。

➢ 接收 RM 的命令并执行。如终止 Container、重启 NM 等。

（3）NodeManager（NM）。NM 是每个集群节点上的资源和任务管理器，整个集群中可以有多个。主要功能如下。

➢ 周期性地向 RM 汇报本节点的资源使用情况和 Container 运行状态。

➢ 接收和处理 AM 发送的 Container 启动/停止等请求。

（4）Container。Container 是 YARN 中对于任务运行资源的抽象，它封装了节点上的各种资源，如 CPU、内存、磁盘、网络等。当 AM 向 RM 申请资源的时候，RM 返回给 AM 的资源便是用 Container 表示的。YARN 会为每一个任务分配一个 Container，任务运行的时候，只能使用该 Container 表示的资源。Container 是一个动态资源划分单位，根据应用程序的需求自动生成。目前，YARN 仅支持 CPU 和内存两种资源。

（5）Client。即客户端，可以提交作业、查询作业的运行进度以及结束作业。

3. YARN 配置常用属性介绍

前面介绍了 YARN 各个组件的功能职责，接下来介绍如何在 Hadoop 平台上配置 YARN。

在 Hadoop 安装完成后，$HADOOP_HOME/etc/hadoop/目录下存放着 YARN 的配置文件 yarn-site.xml，YARN 的配置属性都是在这个 xml 文件中。

（1）RM 相关配置参数

➢ yarn.resourcemanager.hostname：设置 RM 的 hostname。

➢ yarn.resourcemanager.address：设置 RM 对客户端暴露的地址，客户端通过该地址向 RM 提交应用程序，默认值为${yarn.resourcemanager.hostname}:8032。

➢ yarn.resourcemanager.schedule.address：设置 RM 对 AM 暴露的地址，AM 通过地址可以向 RM 申请资源、释放资源，默认值为${yarn.resourcemanager.hostname}:8030。

➢ yarn.resourcemanager.webapp.address：设置 RM 对外暴露的 Web http 地址，用户可以通过该地址在浏览器中查看集群信息和任务执行进度，默认值为${yarn.resourcemanager.hostname}:8088。

➢ yarn.resourcemanager.webapp.https.address：设置 Web https 地址，默认值为${yarn.resourcemanager.hostname}:8090。

➢ yarn.resourcemanager.resource-tracker.address：设置 RM 对 NM 暴露的地址，NM 通过该地址向 RM 汇报心跳、领取任务等，默认值为${yarn.resourcemanager.hostname}:8031。

➢ yarn.resourcemanager.resource-tracker.client.thread-count：设置 RM 处理 NM 的 RPC 请求的 handler 数目，默认值为 50。

➢ yarn.resourcemanager.admin.address：管理员通过该地址向 RM 发送管理命令等，默认值为${yarn.resourcemanager.hostname}:8033。

➢ yarn.resourcemanager.scheduler.class：设置资源调度器主类，默认值为 org.apache.hadoop.yarn.server.resourcemanager.scheduler.capacity.CapacityScheduler。

➢ yarn.resourcemanager.scheduler.client.thread-count：设置 RM 处理 AM 的 RPC 请求的 handler 数目，默认值为 50。

（2）NM 相关配置参数

➢ yarn.nodemanager.local-dirs：设置中间结果存放位置，可以配置多个目录，默认值为${hadoop.tmp.dir}/nm-local-dir。

➢ yarn.nodemanager.aux-services：设置 NM 上运行的附属服务，需要配置成 mapreduce_shuffle，才可以运行 MapReduce 程序。

这里只列举了部分常用属性配置，读者想要了解更多的属性配置请参考官网。

示例 1

在前面搭建的移动通信数据处理平台上配置 YARN，使得可以在 YARN 上执行 MapReduce 程序。YARN 的属性配置请扫描二维码。

YARN 的配置

关键代码：

```
//yarn-site.xml
<property>
    <name> yarn.nodemanager.aux-services </name>
    <value>mapreduce_shuffle</value>
</property>
//mapred-site.xml
<!-- 指定 MapReduce 执行框架为 YARN -->
<property>
    <name>mapreduce.framework.name</name>
    <value>yarn</value>
</property>
```

其他的属性使用默认值即可。

4. 在 YARN 集群上提交作业

前面在 yarn-site.xml 中配置了 YARN 集群环境的属性，接下来在 YARN 上运行 MapReduce 程序。

示例 2

统计移动通信数据文件中每个手机号的通话次数。

测试数据格式如下：

```
1363157985066 13726230503 20180120 60
1363157991076 13926435656 20180221 20
```

1363157985066 13726230503 20180218 5
1363157985066 13726230503 20180218 18
1363157991076 13926435656 20180226 28
1353157991099 13526435699 20180228 8

关键代码：

（1）MapReduce 代码如下所示。

```
import org.apache.commons.lang.StringUtils;
import org.apache.hadoop.conf.Configuration;
import org.apache.hadoop.fs.FileSystem;
import org.apache.hadoop.fs.Path;
import org.apache.hadoop.io.LongWritable;
import org.apache.hadoop.io.Text;
import org.apache.hadoop.mapreduce.Job;
import org.apache.hadoop.mapreduce.Mapper;
import org.apache.hadoop.mapreduce.Reducer;
import org.apache.hadoop.mapreduce.lib.input.FileInputFormat;
import org.apache.hadoop.mapreduce.lib.output.FileOutputFormat;
import org.apache.hadoop.io.IntWritable;
import java.io.IOException;

/**
* 操作 mobile.txt 文件，统计移动通信数据文件中每个手机号的通话次数
/
public class MobileNum {
    private final static String INPUT_PATH = "hdfs://hadoop:8020/mobile/yarn/input";
    private final static String OUTPUT_PATH =
                                    "hdfs://hadoop:8020/mobile/yarn/output";
    public static class MobileMapper    extends Mapper<LongWritable,Text, Text,IntWritable>{
        public static final IntWritable one = new IntWritable(1);
        private Text phone = new Text();
        public void map(LongWritable key, Text value,Context context)throws IOException,
        InterruptedException {
            String line = value.toString();
            String[] fields = StringUtils.split(line, " ");
            String phoneNB = fields[1];
            phone.set(phoneNB);
            context.write(phone,one);
        }
    }
    public static class FlowSumReducer extends Reducer<Text, IntWritable, Text, IntWritable> {
        public void reduce(Text key, Iterable<IntWritable> value, Context context) throws
        IOException,InterruptedException {
```

```
            int counter=0;
            for(IntWritable i : value){
                counter+=i.get();
            }
            context.write(key, new IntWritable(counter));
        }
    }
    public static void main(String[] args) throws IOException,   ClassNotFoundException,
    InterruptedException {
        //得到 hadoop 的一个配置参数
        Configuration conf = new Configuration();
        //获取一个 job 实例
        Job job = Job.getInstance(conf);
        //加载 job 的运行类
        job.setJarByClass(MobileNum.class);
        job.setMapperClass(MobileMapper.class);
        job.setReducerClass(FlowSumReducer.class);
        //设置 mapper 类的输出类型
        job.setOutputKeyClass(Text.class);
        job.setOutputValueClass(IntWritable.class);
        //设置文件的输入路径
        FileInputFormat.setInputPaths(job, new Path(INPUT_PATH));
        //设置文件的输出路径
        FileSystem fs = FileSystem.get(conf);
        Path path = new Path(OUTPUT_PATH);
        if (fs.isDirectory(path)) {
            fs.delete(path, true);
        }
        FileOutputFormat.setOutputPath(job, new Path(OUTPUT_PATH));
        boolean res = job.waitForCompletion(true);
        System.exit(res?1:0);
    }
}
```

（2）使用 mvn clean package -DskipTests 打包，然后上传到/home/hadoop/lib 目录下。

（3）将测试数据上传到 HDFS 目录下。

```
[hadoop@hadoop ~]# hdfs dfs -mkdir -p /mobile/yarn/input
[hadoop@hadoop ~]# hdfs dfs -put mobile.txt /mobile/yarn/input
```

（4）提交作业到 YARN 集群上执行。

```
[hadoop@hadoop ~]# yarn jar mobileOnYarn.jar
```

（5）通过 webui 的方式查看运行状况。YARN 的 webui 界面访问地址为：http://hadoop:8088。界面效果如图 4.3 所示。

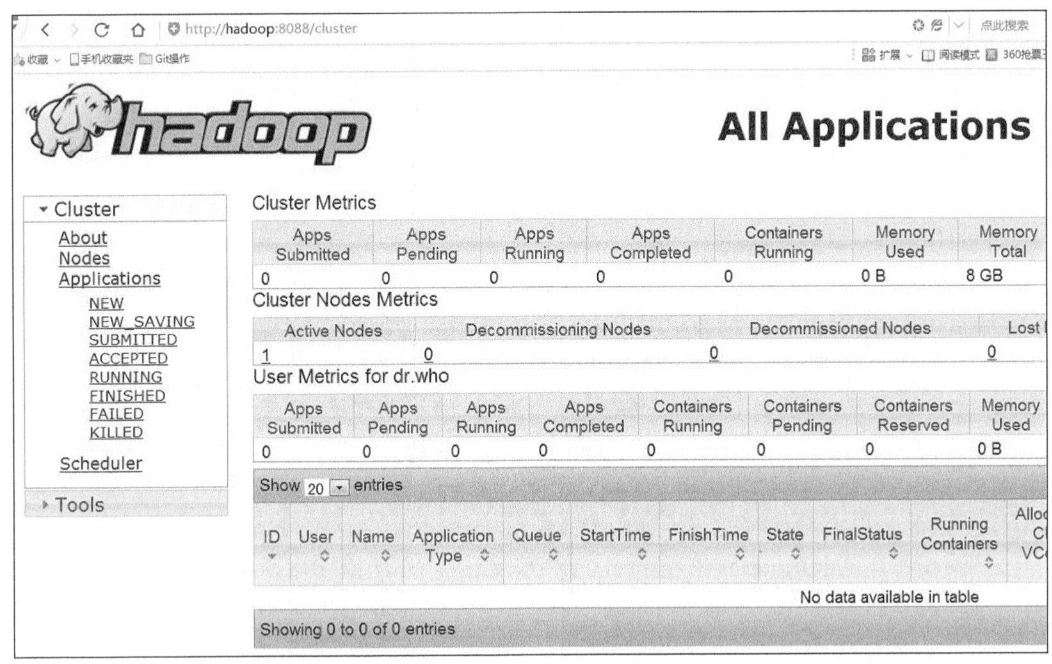

图4.3　YARN的webui管理界面

（6）使用 HDFS 命令查看输出结果。

输出结果：

```
[hadoop@hadoop ~]# hdfs dfs -cat /mobile/yarn/output/part-r-00000
13526435699     1
13726230503     2
13926435656     3
```

4.1.3　YARN 架构设计

1. YARN 的执行流程

YARN 应用程序的生存期跨度比较大，主要分为短应用程序和长应用程序。短应用程序是指一定时间内能运行完成并退出的应用程序，如 MapReduce 作业、Spark 作业。长应用程序是指永不终止运行的应用程序，如 Servie、HttpServer 等，通常用于提供一些服务。尽管两类应用程序的作业不同，但是在 YARN 的执行流程是相同的。YARN 的执行流程如图 4.4 所示。

当用户向 YARN 中提交应用程序后，YARN 将分两个阶段运行该应用程序，第一个阶段是启动 AM（Application Master），第二个阶段是由 AM 创建应用程序，为任务申请资源，并监控任务的整个运行过程，直到任务运行完成。具体的执行流程如下。

（1）Client 向 YARN 中提交应用程序，并请求 RM 分配资源。

（2）RM 开启一个 Container，在 Container 中运行 AM，并与对应的 NM 通信，启动该应用程序的 AM。

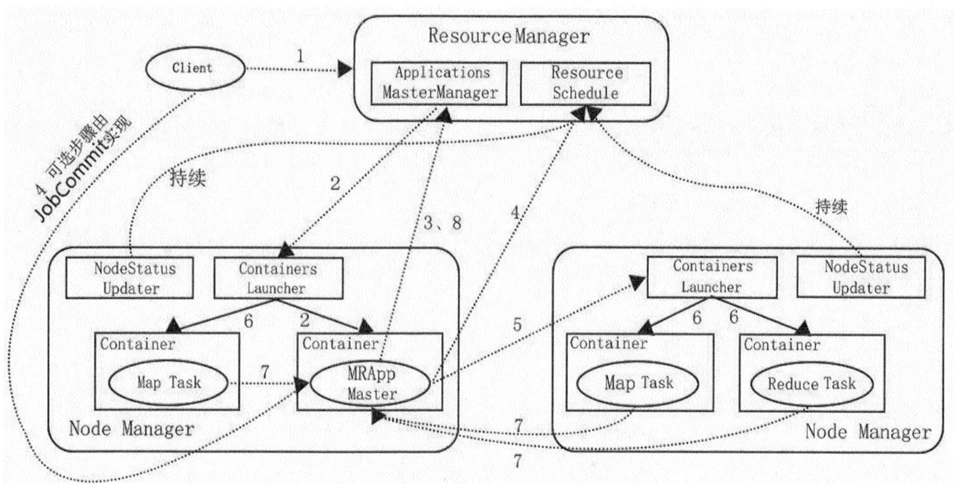

图4.4　YARN的执行流程图

（3）AM 向 Applications Manager（ASM）注册，注册成功后，用户可以通过 RM 查看该应用程序的运行状态，然后为各个任务申请资源，并监控它们的运行状态，直到运行结束。即重复步骤（4）～（7）。

（4）AM 采用轮询的方式通过 RPC 请求方式向 RM 申请和获取资源。

（5）Applications Manager 将资源封装发给 AM，AM 与对应的 NM 通信，在 NM 上启动任务。

（6）NM 启动任务。

（7）各个任务通过 RPC 请求的方式向 AM 汇报本任务的状态和进度，让 AM 可以随时掌握各个任务的运行状态，确保任务失败时可以重新启动任务。AM 将任务的最终执行结果反馈给 Applications Manager。

（8）应用程序执行完成后，AM 向 RM 注销并关闭自己。

2. YARN 的资源管理

在 YARN 架构中，资源调度和隔离是最重要且最基础的两个功能。资源通常就是指内存、CPU、IO 三种资源。到目前为止，YARN 仅支持 CPU 和内存两种资源的管理和调度。前面介绍 YARN 的配置时，没有对资源管理的参数进行配置，采用的都是默认的参数。本节主要针对资源管理的参数配置进行详细分析。

（1）YARN 的内存管理

YARN 允许用户配置每个节点上可用的物理内存资源，具体配置参数如下。

➢ yarn.nodemanager.resource.memory-mb：表示该节点上可以使用的物理内存总量，默认是 8129。

➢ yarn.nodemanager.vmem-pmem-ratio：表示使用 1MB 物理内存最多可以使用的虚拟内存量，默认是 2.1。

➢ yarn.nodemanager.pmem-check-enabled：表示是否启用一个线程来检查每个任务正在使用的物理内存量，如果超出了分配值，则直接将其 kill，默认是 true。

➢ yarn.schedule.minimum-allocation-mb：表示单个任务可以使用的最小物理内存量，默认是 1024。

➢ yarn.schedule.maximum-allocation-mb：表示单个任务可以申请的最大物理内存量，默认是 8192。

（2）YARN 的 CPU 管理

在 YARN 中，和 CPU 相关的配置参数如下。

➢ yarn.nodemanager.resource.cpu-vcores：表示该节点上 YARN 可使用的虚拟 CPU 个数，默认是 8。

➢ yarn.schedule.minimum-allocation-vcores：表示单个任务可申请最小 CPU 个数，默认是 1。

➢ yarn.schedule.maximum-allocation-vcores：表示单个任务可申请最大 CPU 个数，默认是 32。

3．YARN 的容错机制

在实际应用中，进程崩溃、机器故障等均容易造成 YARN 任务执行失败。YARN 的好处之一就是提供了针对此类故障的容错机制，保证能够成功完成作业。

（1）RM 容错。由于集群中只有一个 RM，所以存在单点故障。目前，解决方式是基于 ZooKeeper 实现高可用（High Available，HA）。也就是提供一个 RM 的备份节点，当主节点出现故障时，将切换到从节点继续工作。

（2）NM 容错。如果 NM 在一定时间内未向 RM 汇报心跳信息，则 RM 认为该 NM 已经死掉了，RM 会将失败任务告诉对应的 AM，由 AM 决定如何处理失败的任务。

（3）AM 容错。不同的应用程序对应不同的 AM，RM 负责监控 AM 的运行状态，一旦 AM 运行失败或超时，RM 将为该 AM 重新分配资源并启动。启动后，AM 需要处理内部任务的容错问题，并且保存已经运行完成的任务，重启时，已经完成的任务无须重新运行。

（4）Container 容错。如果 AM 在一定时间内未启动分配到的 Container，则 RM 会将该 Container 状态设置为失败并回收；如果一个 Container 在运行过程中因为外界因素导致运行失败，则 RM 会转告对应的 AM，由它决定如何处理。

4．YARN 的设计目标

YARN 被设计成通用的、统一的资源管理系统。目前，在 YARN 上不仅可以运行短应用程序，而且可以运行作为服务提供的长应用程序。同时，在 YARN 上可以运行各种不同框架的作业。比如：MapReduce（离线计算框架）、Tez（DAG 计算框架）、Storm（流式计算框架）、Spark（内存计算框架）。

4.1.4　技能实训

自己动手完成移动通信数据处理平台的 YARN 集群环境配置，并在 YARN 集群上提交第 3 章编写的 WorldCount 应用程序。

任务 2 配置 YARN 容错

【任务描述】

掌握 YARN 架构的容错配置，包括 ResourceManager 自动重启和高可用机制。

【关键步骤】

（1）配置 ResourceManager 自动重启。

（2）配置 ResourceManager 高可用。

4.2.1 ResourceManager 自动重启

1. ResourceMananger 自动重启阶段

在 YARN 架构设计中，只存在一个 ResourceManager，因此在 YARN 集群中可能存在单点故障，这就要求当 ResourceManager 出现故障时，需要尽快重启 ResourceManager，以尽可能地减少损失，并且保证 ResourceManager 的重启过程对最终用户不可见。YARN 的重启可以分为两个阶段。

（1）第一阶段：增强的 ResourceManager 将应用程序的状态和其他认证信息保存到一个插入式的状态存储中，ResourceManager 重启时将从状态存储中重新加载这些信息，重新开始之前正在运行的应用程序，这样可以保证用户不用重新提交自己的应用程序。这个阶段的功能在 Hadoop-2.4.0 版本已经实现。

（2）第二阶段：重启时通过从 NodeManager 读取容器的状态，以及从 ApplicationMaster 读取容器的请求，重构 ResourceManager 的运行状态。在此阶段中，之前正在运行的应用程序将不会在 ResourceManager 重启后被杀死，所以应用程序不会因为 ResourceManager 中断而丢失工作。这个阶段的功能在 Hadoop-2.6.0 版本已经实现。

2. 配置 ResourceManager 自动重启

学习了 ResourceManager 自动重启的特性和机制后，接下来就是如何启用 ResourceManager 重启的特性，主要是在 YARN 的配置文件 yarn-site.xml 中配置如下参数属性。

```
//yarn-site.xml
#开启 ResourceManager 重启功能，默认为 false
<property>
    <name>yarn.resourcemanager.recovery.enabled</name>
    <value>true</value>
</property>
#用于状态存储的类，有三种 StateStore，默认是 FileSystemRMStateStore
<property>
    <name>yarn.resourcemanager.store.class</name>
    <value>org.apache.hadoop.yarn.server.resourcemanager.recovery.ZKRMStateStore</value>
</property>
```

```
#被 ResourceManager 用于状态存储的 ZooKeeper 服务器地址，多个地址之间使用逗号分隔
<property>
    <name>yarn.resourcemanager.zk-address</name>
    <value>hadoop:2181</value>
</property>
```

ZooKeeper 的安装部署会在下一章进行介绍，这里就不再进行详细分析，读者知道如何进行配置即可。完成 ZooKeeper 的安装部署后，只需要将自己安装的 ZooKeeper 地址配置在配置文件中对应的位置上就可以完成 ResourceManager 的自动重启配置。

4.2.2　ResourceManager 高可用

针对 ResourceManager 的单点故障问题，YARN 不仅提供了启动重启的方式，同时也提供了 ResourceManager 的高可用机制。在 Hadoop-2.4.0 版本中，添加了 Active/Standby（主动/备用）ResourceManager 的方式。架构如图 4.5 所示。

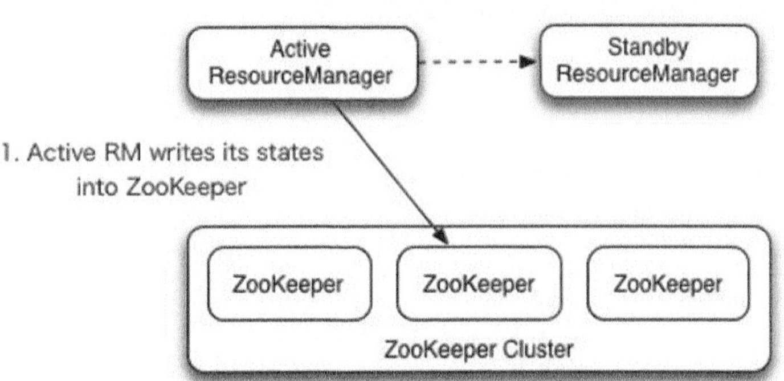

图4.5　ResourceManager高可用机制

ResourceManager 的高可用是通过 Active/Standby 架构模式实现的，Active ResourceManager 会将状态信息写入到 ZooKeeper 集群中，如果 Active ResourceManager 宕机，基于 ZooKeeper 保存的状态信息，可以实现将 Standby ResourceManager 切换成 Active ResourceManager。完成切换的方式有手动切换和自动切换两种。

（1）手动切换。当未启用自动切换时，管理员必须手动将其中一个 ResourceManager 转换为 Active。命令如下。

```
#查看当前 RM 的状态
yarn rmadmin -getServieState rm1
#手动切换 ResourceManager
yarn rmadmin -transitionToStandby rm1
```

手动切换的最小需求配置项如下。在 YARN 的配置文件 yarn-site.xml 中进行配置。

```
<!--开启 yarn 的高可用-->
<property>
    <name>yarn.resourcemanager.ha.enabled</name>
```

```
        <value>true</value>
    </property>
    <!-- 指定 ResourceManager 主备列表 -->
    <property>
        <name>yarn.resourcemanager.ha.rm-ids</name>
        <value>rm1,rm2</value>
    </property>
    <!-- 分别指定 ResourceManager 地址 -->
    <property>
        <name>yarn.resourcemanager.hostname.rm1</name>
        <value>hadoop</value>
    </property>
    <property>
        <name>yarn.resourcemanager.webapp.address.rm1</name>
        <value>hadoop:8088</value>
    </property>
    <property>
        <name>yarn.resourcemanager.hostname.rm2</name>
        <value>hadoop01</value>
    </property>
    <property>
        <name>yarn.resourcemanager.webapp.address.rm2</name>
        <value>hadoop01:8088</value>
    </property>
    <!--指定 ResourceManager 的 cluster id-->
    <property>
        <name>yarn.resourcemanager.cluster-id</name>
        <value>yarn-ha</value>
    </property>
    <!--关闭自动故障转移-->
    <property>
        <name>yarn.resourcemanager.ha.automatic-failover.enabled</name>
        <value>false</value>
    </property>
```

这里的配置采用的是两个节点的 ResourceManager。一个 Active 节点，一个 Standby 节点。要实现高可用，需要参照第 1 章介绍的 Hadoop 平台的安装步骤，为 Hadoop 平台添加一个节点 hadoop01。

注意

如果启用了自动故障转移，则不能使用手动切换命令。当 yarn.resourcemanager.ha.enabled 属性值为 true 时，默认开启自动故障转移功能，也就是说 yarn.resourcemanager.ha.automatic-failover.enabled 的值为 true。

（2）自动切换。ResourceManager 可以选择基于 ZooKeeper 的 ActiveStandbyElector 来决定哪个 ResourceManager 应该是活动的。当 Active ResourceManager 出现故障时，另一个 ResourceManager 被自动选择为 Active ResourceManager。

自动切换的配置是在 YARN 的配置文件 yarn-site.xml 中进行的。配置属性如下。

```
<!-- 开启 ResourceManager 高可用 -->
<property>
    <name>yarn.resourcemanager.ha.enabled</name>
    <value>true</value>
</property>
<!--指定 ResourceManager 的 cluster id -->
<property>
    <name>yarn.resourcemanager.cluster-id</name>
    <value>yarn-ha</value>
</property>
<!-- 指定 ResourceManager 主备列表 -->
<property>
    <name>yarn.resourcemanager.ha.rm-ids</name>
    <value>rm1,rm2</value>
</property>
<!-- 分别指定 ResourceManager 地址 -->
<property>
    <name>yarn.resourcemanager.hostname.rm1</name>
    <value>hadoop</value>
</property>
<property>
    <name>yarn.resourcemanager.webapp.address.rm1</name>
    <value>hadoop:8088</value>
</property>
<property>
    <name>yarn.resourcemanager.hostname.rm2</name>
    <value>hadoop01</value>
</property>
<property>
    <name>yarn.resourcemanager.webapp.address.rm2</name>
    <value>hadoop01:8088</value>
</property>
<!-- 指定 ZooKeeper 集群地址，多个地址之间用逗号分隔 -->
<property>
    <name>yarn.resourcemanager.zk-address</name>
    <value>hadoop:2181</value>
</property>
```

ResourceManager 的高可用自动切换需要使用 ZooKeeper 来完成，读者在学习完第 5 章的 ZooKeeper 安装部署后，将自己的 ZooKeeper 集群地址配置在上述配置文件中对应的位置即可完成 ResourceManager 的高可用配置。

更多有关 YARN HA 的配置原理请扫描二维码。

YARN HA

本章总结

- YARN 是 Hadoop 的核心构成，它是一个通用的资源管理系统，在 YARN 上不仅可以运行 MapReduce 作业，也可以运行 Spark、Storm 等作业。
- YARN 是典型的 Master/Slave 架构。在一个 YARN 集群中，包含一个活动的 ResourceManager 和多个 NodeManager。
- YARN 提供了两种解决 ResourceManager 单点故障问题的方式，分别是 ResourceManager 自动重启和 ResourceManager 高可用机制，这两种方式都可以在 yarn-site.xml 配置文件中通过属性设置进行配置。

本章作业

一、简答题

1．YARN 的核心组件有哪些？每个组件的具体功能是什么？

2．YARN 架构与 MapReduce1.0 架构相比的优势是什么？

二、编码题

1. 将自己编写的第 3 章中任务 1 的技能实训：统计用户手机流量的功能代码提交到 YARN 集群运行并查看运行结果。具体步骤参考本章示例 2。

2. 将自己编写的第 3 章中任务 2 的技能实训：查找微信共同好友列表的功能代码提交到 YARN 集群运行并查看运行结果。具体步骤参考本章示例 2。

3. 将自己编写的第 3 章中任务 3 的技能实训：自定义计数器的功能代码提交到 YARN 集群运行并查看运行结果。具体步骤参考本章示例 2。

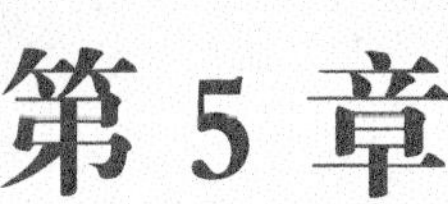

第5章 ZooKeeper 简介及安装

技能目标

- 理解 ZooKeeper 的作用
- 掌握 ZooKeeper 的架构设计
- 掌握 ZooKeeper 的数据模型
- 学会搭建 ZooKeeper 单机环境

本章任务

任务1　了解 ZooKeeper

任务2　搭建 ZooKeeper 单机环境

任务3　实现分布式系统服务器上下线的动态感知

本章资源下载

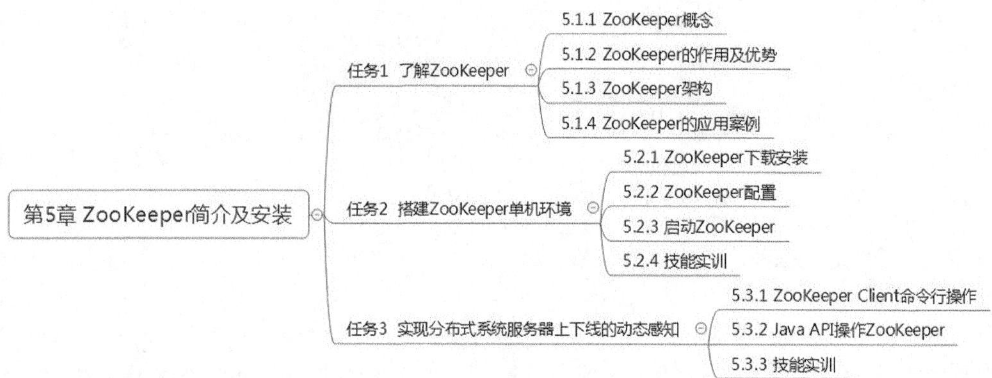

ZooKeeper 是一个分布式的、开放源码的分布式应用程序协调服务，是 Hadoop 的一个子项目，是对 Google 分布式同步系统 Chubby 的开源实现。本章主要介绍 ZooKeeper 架构、ZooKeeper 的作用、搭建 ZooKeeper 单机环境并在搭建的环境上实现其 Client 端操作和 Java API 操作。

任务 1 了解 ZooKeeper

【任务描述】

理解 ZooKeeper 架构设计以及 ZooKeeper 的作用及优势，了解 ZooKeeper 的应用场景。

【关键步骤】

（1）了解 ZooKeeper 是什么以及它的作用和优势。

（2）理解 ZooKeeper 架构设计。

（3）了解 ZooKeeper 的实际应用。

5.1.1 ZooKeeper 概念

1．ZooKeeper 概述

ZooKeeper 是一个为分布式应用提供一致性服务的软件，针对大型分布式系统设计，主要用来解决实际分布式应用场景中存在的一些问题，提供配置维护、名字服务、分布式同步、组服务等功能，在分布式模式下，能够为分布式应用提供高性能和高可靠的协调服务，极大的降低了开发分布式应用的成本。

2．ZooKeeper 的设计目标

ZooKeeper 的设计目标主要包括以下几点。

（1）简单（simple）。分布式应用中的各个进程可以通过 ZooKeeper 的名字空间（Namespace）来协调，这个名字空间是共享的、具有层次结构的，最重要的是它类似于平常接触到的文件系统的目录结构，非常简单。

（2）可复制（replicated）。在 ZooKeeper 集群中，每个节点的数据都可以在集群中复制传播，实现集群中每个节点的数据同步一致，同时也可以避免发生单点故障。

（3）顺序性（order）。ZooKeeper 使用时间戳来记录引起状态更新的事务性操作并通过时间戳来保证有序性。顺序性包括全局有序和偏序两种，全局有序是指如果一台服务器上的消息 a 在消息 b 前发布，则在所有服务器上消息 a 都将在消息 b 前被发布。偏序是指如果一个消息 b 在消息 a 后被同一个发送者发布，则消息 a 必将排在消息 b 前。基于这一特性，ZooKeeper 可以实现更高级的抽象操作，比如同步等。

（4）快速（fast）。在分布式应用中，ZooKeeper 对于以读数据为主的应用场景，是非常高效的。

5.1.2 ZooKeeper 的作用及优势

1. 为什么需要 ZooKeeper

为什么需要 ZooKeeper 呢？主要表现在如下几个方面。

（1）开发分布式系统是非常复杂的。复杂主要体现在集群的维护和多节点应用程序的协作运行上，涉及的细节问题非常多，比如同一个配置在多台机器上的同步、客户端程序实时感知服务器状态、应用程序之间的公共资源的互斥访问等问题。这些问题如果都依靠开发人员和维护人员去解决，将非常耗费人力，也达不到实时准确的效果。这就需要开发一个协调程序来专门解决这样的问题。

（2）开发的私有的协调程序间缺乏一个通用的机制，而且由于反复编写，难以形成通用的、伸缩性好的协调程序。

ZooKeeper 能很好的解决上述两个问题，ZooKeeper 天生就是为解决分布式协调服务这个问题而来的。

2. ZooKeeper 作用

作为一个分布式协调服务框架，ZooKeeper 可以实现的功能包括：配置管理、分布式同步以及集群管理等。下面分别介绍 ZooKeeper 在实现这些功能方面的优势。

（1）配置管理。在实际开发应用的过程中，除了编写代码之外，还有一些配置工作，比如数据库连接，通常都是使用配置文件来完成的。在分布式系统中，往往很多服务都依赖于这些配置文件，这就可用 ZooKeeper 来提供统一的配置服务。

（2）分布式同步。ZooKeeper 提供了锁服务，用来解决分布式系统操作过程中的数据一致性问题，以保证分布式数据运算过程中数据操作的一致性。

（3）在分布式的集群中，节点故障管理是非常麻烦的一件事情，很多大数据技术框架都存在单点故障的问题，利用 ZooKeeper 的通信与回调机制可以进行分布式集群的机器状态监视，做到动态感知集群目前的状态，同时实现很多技术框架的主备切换。

5.1.3 ZooKeeper 架构

1. ZooKeeper 架构

ZooKeeper 集群由一组 Server 节点和若干 Client（客户端）组成，这一组 Server 节点中存在一个角色为 Leader 的节点，其他节点为 Follower，Client（客户端）为请求发起方。ZooKeeper 基本架构如图 5.1 所示。

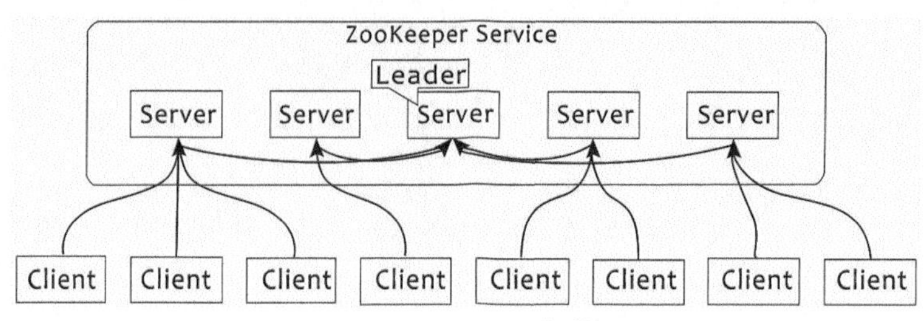

图5.1　ZooKeeper架构图

2. ZooKeeper 工作原理

ZooKeepe 架构分为客户端和服务器端，客户端需要连接到整个 ZooKeeper 集群的某个服务器上，使用并维护一个 TCP 连接，通过该连接发送请求、接收响应、获取观察的事件以及发送心跳。

启动 ZooKeeper 集群后，多个 ZooKeeper 服务器会选举出一个 Leader，选举 Leader 的目的就是为了保证分布式环境中数据的一致性。

当客户端执行写操作时，所有的写请求都会先被转发到 Leader 上，Leader 再将更新操作广播到 Follower。Leader 在接收到数据变更请求后，首先将变更写入本地磁盘，方便恢复使用。当大多数的节点持久化存储这个更新后，Leader 会提交更新，并回复客户端写操作成功。

ZooKeeper 原理

3. ZooKeeper 数据模型

（1）ZooKeeper 名字空间。ZooKeeper 拥有一个层次的名字空间，和标准的文件系统的层次树结构非常类似，名字是一系列以“/”隔开的路径元素。ZooKeeper 名字空间中所有的节点都是通过路径识别的。与标准的文件系统不同的是，ZooKeeper 名字空间中的每个节点可以有数据也可以有子目录，ZooKeeper 的数据节点被称为 znode。ZooKeeper 的树结构如图 5.2 所示。

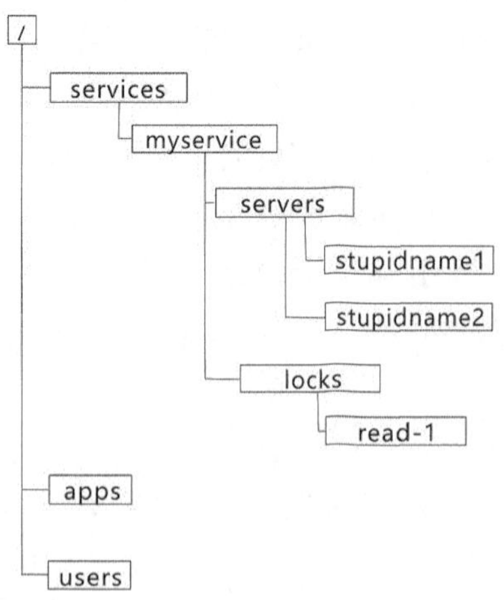

图5.2　ZooKeeper树结构图

（2）znode 结构。ZooKeeper 名字空间中的 znode 兼具文件和目录两种特点，既像文件一样维护着数据、元数据信息、ACL（访问控制列表）信息、时间戳等数据结构，又像目录一样可以作为路径标识的一部分。每个 znode 包含以下三个部分。

- stat。状态信息，描述 znode 的版本、权限等信息。
- data。znode 关联的数据。
- children。znode 下的子节点。

ZooKeeper 中的 znode 有两种类型：临时节点（Ephemeral）和永久节点（Persistent）。节点的类型要在创建的时候指定，并且不能修改。

- 临时节点不是持久节点，一旦与客户端的会话结束，节点会自动删除，也就是说，临时节点的生命周期依赖于创建它们的会话。值得注意的是，临时节点不允许拥有子节点。
- 永久节点的生命周期不依赖于创建它们的会话，只有在客户端执行删除操作的时候，永久节点才会被删除。

ZooKeeper 中有关 znode 的数据变更使用 Zxid（ZooKeeper Transaction Id）来表示，一个 Zxid 与一个时间戳相对应，所以多个不同的变更对应的事务是有序的。

（3）Watch（监视器）。客户端可以在 znode 上设置监视器，设置了监视器后，znode 发生变化时，客户端可以接收到通知。ZooKeeper 中的监视器只能触发一次，也就是说，当 znode 第二次发生变化时，如果没有重新设置 znode 的监视器，ZooKeeper 将不会给客户端发送变更通知。目前，znode 可以设置两种监视器，分别为 Data Watch（znode 发生数据变更时触发的监视器事件）和 Child Watch（znode 子节点发生变更时触发的监视器事件）。

（4）序列节点（Sequence Node）。创建 znode 时，可以请求 ZooKeeper 生成序列。序列以路径名为前缀，计数器紧跟在路径后面，比如，可以生成如下序列形式：x-0000000001、x-0000000002、x-0000000003。对于 znode 的父节点来说，序列中的每个计数器字符串都是唯一的。

5.1.4 ZooKeeper 的应用案例

（1）ZooKeeper 在 HBase 中的使用。HBase 内置了 ZooKeeper 来完成集群的协调。使用 ZooKeeper，HBase 集群中可以启动多个 HMaster（HBase 的主节点）节点，并且保证只有一个 HMaster 在运行。

（2）ZooKeeper 在 Hadoop 集群中的使用。前面已经介绍过，在 Hadoop 集群中只能有一个活动的 NameNode，为了避免单点故障，通常都要配置 NameNode 的高可用机制，NameNode 的高可用就是使用 ZooKeeper 来实现的。

（3）ZooKeeper 在 YARN 高可用中的使用。第 4 章曾介绍过 YARN 自动重启机制和高可用机制都是使用 ZooKeeper 来实现的。

任务 2 搭建 ZooKeeper 单机环境

【任务描述】

搭建 ZooKeeper 单机环境。

【关键步骤】

（1）下载安装 ZooKeeper。

（2）配置 ZooKeeper。

（3）启动 ZooKeeper 并验证。

5.2.1 ZooKeeper 下载安装

第 1 章介绍过 Hadoop 平台的安装，ZooKeeper 的安装方式与 Hadoop 平台的安装方式十分类似，本节主要介绍 ZooKeeper 的单机环境的搭建。

ZooKeeper
安装

1. 安装包下载

本书采用的是 CDH 版本的 ZooKeeper-3.4.5-cdh5.14.2，读者可以去官网下载对应的版本。软件包下载完成后，可以通过远程工具上传到第 1 章创建的虚拟机服务器上，放在/home/hadoop/software 目录下。

2. 解压安装

（1）下载完成以后，需要将安装包解压。

```
[hadoop@hadoop ~]$ tar –zxvf /home/hadoop/software/zookeeper-3.4.5-cdh5.14.2.tar.gz
```

（2）将解压后的文件复制到/opt 目录下。

```
[hadoop@hadoop ~]$ sudo mv /home/hadoop/software/zookeeper-3.4.5-cdh5.14.2
/opt/zookeeper-3.4.5-cdh5.14.2
```

（3）配置环境变量。

```
[hadoop@hadoop ~]$ vi ~/.bashrc
```

在打开的文件中，添加如下两行：

```
export ZK_HOME=/opt/zookeeper-3.4.5-cdh5.14.2
export PATH=$ZK_HOME/bin:$PATH
```

使配置文件生效。

```
[hadoop@hadoop ~]$ source ~/.bashrc
```

5.2.2 ZooKeeper 配置

ZooKeeper 解压安装并配置环境变量后，前置工作就完成了，接下来对 ZooKeeper 的配置文件进行配置。

1. 配置 zoo.cfg 文件

ZooKeeper 主配置文件 zoo.cfg 在$ZK_HOME/conf/目录下，默认是没有该文件的，但是提供了 zoo_sample.cfg 模板文件。需要执行下面的命令来创建 zoo.cfg 文件。

```
[hadoop@hadoop ~]$ cp –a /opt/zookeeper-3.4.5-cdh5.14.2/conf/zoo_sample.cfg  /opt/zookeeper-
```

3.4.5-cdh5.14.2/conf/zoo.cfg

创建完成后，需要配置 zoo.cfg 文件，主要配置 ZooKeeper 数据文件目录 dataDir。由于 dataDir 的默认值是在/tmp 目录下，系统重启后会清空数据，所以需要将文件中 dataDir 的值设置为自己的本地路径，步骤如下。

（1）创建数据文件目录。

```
[hadoop@hadoop ~]$ sudo mkdir /opt/ zookeeper-3.4.5-cdh5.14.2/zkData
```

（2）配置 dataDir。

```
[hadoop@hadoop ~]$ vi /opt/zookeeper-3.4.5-cdh5.14.2/conf/zoo.cfg
```

将 dataDir 的值修改为/opt/ zookeeper-3.4.5-cdh5.14.2/zkData，对应代码如下。

```
dataDir=/opt/ zookeeper-3.4.5-cdh5.14.2/zkData
```

2. **配置 ZooKeeper 的日志路径**

（1）创建日志存放目录。

```
[hadoop@hadoop ~]$ sudo mkdir /opt/ zookeeper-3.4.5-cdh5.14.2/logs
```

（2）配置 ZOO_LOG_DIR。

```
[hadoop@hadoop ~]$ vi /opt/zookeeper-3.4.5-cdh5.14.2/libexec/zkEnv.sh
```

将 ZOO_LOG_DIR 的值修改为/opt/ zookeeper-3.4.5-cdh5.14.2/logs，对应代码如下。

```
ZOO_LOG_DIR=/opt/ zookeeper-3.4.5-cdh5.14.2/logs
```

5.2.3 启动 ZooKeeper

ZooKeeper 配置完成以后，就可以启动了。与 ZooKeeper 启动和停止相关的命令都放在$ZK_HOME/bin 目录下。

1. **启动 ZooKeeper**

（1）启动 ZooKeeper。

```
[hadoop@hadoop ~]$ zkServer.sh start
```

（2）停止 ZooKeeper。

```
[hadoop@hadoop ~]$ zkServer.sh stop
```

2. **查看 ZooKeeper 运行模式**

```
[hadoop@hadoop ~]$ zkServer.sh status
Mode: standalone
```

3. **验证**

Hadoop 启动成功以后，通常有以下几步操作。

（1）通过 jps 命令查看启动的进程。

```
[hadoop@hadoop ~]$ jps
26471 QuorumPeerMain
```

可以看到进程 QuorumPeerMain，说明 ZooKeeper 已经安装完并启动成功了。

（2）使用 ZooKeeper 客户端验证连接。

```
[hadoop@hadoop ~]$ zkCli.sh –server localhost:2181
```

5.2.4 技能实训

按照任务 2 的步骤，读者搭建自己的 ZooKeeper 单机环境。

关键步骤：

（1）下载对应版本的安装包安装并配置环境变量。

（2）修改 ZooKeeper 中的配置文件，包括修改数据文件目录和日志目录的值。

（3）启动 ZooKeeper 并验证是否启动成功。

任务 3 实现分布式系统服务器上下线的动态感知

【任务描述】

掌握 ZooKeeper Client 和 Java API 操作，并使用 ZooKeeper Java API 实现分布式系统服务器上下线的动态感知。

【关键步骤】

（1）使用 Client 命令行操作 ZooKeeper。

（2）使用 Java API 操作 ZooKeeper。

（3）实现分布式系统服务器上下线的动态感知。

5.3.1 ZooKeeper Client 命令行操作

ZooKeeper 为用户提供了 Client 操作命令来管理 ZooKeeper 上的数据。

1. 连接 ZooKeeper 服务

【命令】

```
zkCli.sh -server host:port
```

示例 1

连接 ZooKeeper 服务。

```
[hadoop@hadoop ~]$ zkCli.sh  -server localhost:2181
```

2. 使用 ZooKeeper 客户端命令完成服务器列表的管理

（1）查看目录下节点

【命令】

```
ls path
```

其中，path 为节点路径。

示例 2

查看 ZooKeeper 根节点下的所有节点。

```
[zk: 127.0.0.1:2181(CONNECTED) 0] ls /
[zookeeper]
```

（2）创建节点并设置值

【命令】

```
create [-s] [-e]  path  data  acl
```

其中，-s 和-e 分别指节点特性：永久或临时节点。默认情况下，不添加-s 或-e 参数，创建的是永久节点。

acl：用来进行权限控制，默认情况下，不做任何权限控制。

path：节点路径。

data：节点保存的数据。

示例 3

在根目录下创建服务器列表目录 serverZnode 并设置值为 127.0.0.1。

```
[zk: 127.0.0.1:2181(CONNECTED) 1] create /serverZnode 127.0.0.1
```

查看是否创建成功。

```
[zk: 127.0.0.1:2181(CONNECTED) 2] ls /
[zookeeper,serverZnode]
```

（3）查看节点内容

【命令】

```
get path
```

示例 4

查看示例 3 的服务器列表内容。

```
[zk: 127.0.0.1:2181(CONNECTED) 3] get /serverZnode
127.0.0.1
cZxid = 0x67
ctime = Thu Jul 05 11:41:13 CST 2018
mZxid = 0x67
mtime = Thu Jul 05 11:41:13 CST 2018
pZxid = 0x67
cversion = 0
dataVersion = 0
aclVersion = 0
ephemeralOwner = 0x0
dataLength = 9
numChildren = 0
```

（4）修改节点的内容

【命令】

```
set path data
```

示例 5

修改服务器列表节点的值。

```
[zk: 127.0.0.1:2181(CONNECTED) 4] set /serverZnode 192.168.85.239
cZxid = 0x67
ctime = Thu Jul 05 11:41:13 CST 2018
mZxid = 0x68
mtime = Thu Jul 05 11:43:31 CST 2018
pZxid = 0x67
cversion = 0
dataVersion = 1
aclVersion = 0
ephemeralOwner = 0x0
```

```
dataLength = 14
numChildren = 0
```

查看是否修改成功。

```
[zk: 127.0.0.1:2181(CONNECTED) 5] get /serverZnode
192.168.85.239
cZxid = 0x67
ctime = Thu Jul 05 11:41:13 CST 2018
mZxid = 0x68
mtime = Thu Jul 05 11:43:31 CST 2018
pZxid = 0x67
cversion = 0
dataVersion = 1
aclVersion = 0
ephemeralOwner = 0x0
dataLength = 14
numChildren = 0
```

（5）删除节点

【命令】

```
delete path
```

示例 6

删除服务器列表 serverZnode。

```
[zk: 127.0.0.1:2181(CONNECTED) 6] delete /serverZnode
```

查看 serverZnode 节点是否删除成功。

```
[zk: 127.0.0.1:2181(CONNECTED) 7] ls /
[zookeeper]
```

3. 使用 help 命令获取所有命令用法

示例 7

使用 help 命令查看 ZooKeeper 客户端命令及用法。

```
[zk: 127.0.0.1:2181(CONNECTED) 8] help
ZooKeeper -server host:port cmd args
        stat path [watch]
        set path data [version]
        ls path [watch]
        delquota [-n|-b] path
        ls2 path [watch]
        setAcl path acl
        setquota -n|-b val path
        history
        redo cmdno
        printwatches on|off
        delete path [version]
        sync path
        listquota path
```

```
rmr path
get path [watch]
create [-s] [-e] path data acl
addauth scheme auth
quit
getAcl path
close
connect host:port
```

5.3.2 Java API 操作 ZooKeeper

ZooKeeper 不仅提供了命令行的操作方式，还提供了 Java API 的操作方式。下面介绍如何使用 Java API 对 ZooKeeper 进行操作，完成 5.3.1 节演示的服务器列表的管理操作。

1. 基本操作

在 IDEA 中创建 maven 项目 ZooKeeperAPI。添加 Maven 依赖包。Maven pom 文件如下。

Maven pom 文件：

```
<properties>
    <project.build.sourceEncoding>UTF-8</project.build.sourceEncoding>
    <zookeeper.version>3.4.5-cdh5.14.2</ zookeeper.version>
</properties>
<dependencies>
    <dependency>
        <groupId>org.apache.zookeeper</groupId>
        <artifactId>zookeeper</artifactId>
        <version>${zookeeper.version}</version>
    </dependency>
    <dependency>
        <groupId>org.slf4j</groupId>
        <artifactId>slf4j-log4j12</artifactId>
        <version>1.7.25</version>
    </dependency>
    <dependency>
        <groupId>org.slf4j</groupId>
        <artifactId>slf4j-api</artifactId>
        <version>1.7.25</version>
    </dependency>
    <dependency>
        <groupId>log4j</groupId>
        <artifactId>log4j</artifactId>
        <version>1.2.17</version>
    </dependency>
    <dependency>
```

```
            <groupId>junit</groupId>
            <artifactId>junit</artifactId>
            <version>4.11</version>
        </dependency>
    </dependencies>
```

本节采用单元测试的方式来演示如何使用 Java API 操作 ZooKeeper。在编写 ZooKeeper 测试代码时，初始化 ZooKeeper 实例的操作可以放在 setUp 方法中，而关闭 ZooKeeper 实例的操作可以放在 tearDown 方法中。

示例 8

创建 ZooKeeper 单元测试的 setUp 和 tearDown 方法。

分析：

使用 Java 代码操作 ZooKeeper 服务器可以通过创建 org.apache.zookeeper.ZooKeeper 类的实例对象，然后调用这个类提供的方法和 ZooKeeper 服务器进行交互。

关键代码：

```
import org.apache.zookeeper.CreateMode;
import org.apache.zookeeper.KeeperException;
import org.apache.zookeeper.Watcher;
import org.apache.zookeeper.ZooDefs.Ids;
import org.apache.zookeeper.ZooKeeper;
import org.apache.zookeeper.WatchedEvent;
import org.junit.Test;
import org.junit.Before;
import org.junit.After;
public class ZookeeperAPI {
    //会话超时时间，设置为与默认时间一致
    private static final int SESSION_TIMEOUT = 30000;
    //创建 ZooKeeper 实例
    ZooKeeper zk;
    //创建 Watcher 实例
    Watcher wh = new Watcher() {
        public void process(WatchedEvent event) {
            System.out.println(event.toString());
        }
    };
    @Before
    public void setUp() throws Exception{
        zk = new ZooKeeper("hadoop:2181", ZookeeperAPI.SESSION_TIMEOUT, this.wh);
    }
    @After
    public void tearDown () throws Exception{
        zk.close();
    }
}
```

在初始化 ZooKeeper 实例时，允许接收一个 watcher 参数，此参数将会作为当前客户端的默认 Watcher，主要是响应与链接状态转换有关的事件（比如建立链接、关闭链接等），它不是一次触发即消亡的，而是会伴随整个 ZooKeeper 实例的生命周期。

初始化完成以后，就可以使用创建的 ZooKccpcr 对象来完成服务器列表的管理。

（1）使用 ZooKeeper 实例的 create()方法创建节点

示例 9

在根节点下创建 serverZnode 节点。

关键代码：

```
/**
*创建/serverZnode
*/
@Test
public void createNode(){
    System.out.println("\n 创建 ZooKeeper 节点(znode：serverZnode，数据：127.0.0.1，权限：
                OPEN_ACL_UNSAFE，节点类型：Persistent");
    try {
        zk.create("/serverZnode", "127.0.0.1".getBytes(), Ids.OPEN_ACL_UNSAFE,
                CreateMode.PERSISTENT);
        System.out.println("\n 查看是否创建成功：");
        System.out.println(new String(zk.getData("/serverZnode", false, null)));
    } catch (KeeperException e) {
        e.printStackTrace();
    } catch (InterruptedException e) {
        e.printStackTrace();
    }
}
```

（2）使用 ZooKeeper 实例的 setData()方法修改节点值

示例 10

修改示例 9 创建的/serverZnode 节点的值。

关键代码：

```
/**
*修改/serverZnode 节点的值
*/
@Test
public void setData(){
    System.out.println("\n 修改节点数据");
    try {
        zk.setData("/serverZnode", "192.168.85.239".getBytes(), -1);
    } catch (KeeperException e) {
        e.printStackTrace();
    } catch (InterruptedException e) {
        e.printStackTrace();
```

```
    }
}
```

（3）使用 ZooKeeper 实例的 getData()方法查看节点的值

示例 11

查看示例 10 修改的值是否成功。

关键代码：

```
/**
*查看/serverZnode 节点的值
*/
@Test
public void getData(){
    System.out.println("\n 查看是否修改成功：");
    try {
        System.out.println(new String(zk.getData("/serverZnode", false, null)));
    } catch (KeeperException e) {
        e.printStackTrace();
    } catch (InterruptedException e) {
        e.printStackTrace();
    }
}
```

（4）使用 ZooKeeper 实例的 delete()方法删除节点

示例 12

删除示例 9 创建的/serverZnode 节点。

关键代码：

```
/**
*删除/serverZnode 节点
*/
@Test
public void deleteNode(){
    System.out.println("\n.删除节点");
    try {
        zk.delete("/serverZnode", -1);
    } catch (InterruptedException e) {
        e.printStackTrace();
    } catch (KeeperException e) {
        e.printStackTrace();
    }
}
```

（5）查看节点是否存在

示例 13

查看示例 12 删除的/serverZnode 节点是否还存在。

关键代码：

```
/**
*查看/serverZnode 是否存在
*/
@Test
public void exists(){
    System.out.println("\n6.查看节点是否被删除：");
    try {
        System.out.println("节点状态：[" + zk.exists("/serverZnode", false) + "]");
    } catch (KeeperException e) {
        e.printStackTrace();
    } catch (InterruptedException e) {
        e.printStackTrace();
    }
}
```

2. **高级操作**

ZooKeeper 可以看作是一个具有高可用的文件系统，但这个文件系统没有文件和目录，而是统一使用 znode 的概念，znode 既可以作为保存数据的容器（如同文件），也可以作为保存其他 znode 的容器（如同目录）。所有的 znode 构成一个层次化的名字空间。利用这种层次结构可以建立组与组成员，创建一个以组名为节点名的 znode 作为父节点，然后以组成员名为节点名来创建子节点的 znode。如图 5.3 所示，给出了一组具有层次化结构的 znode。

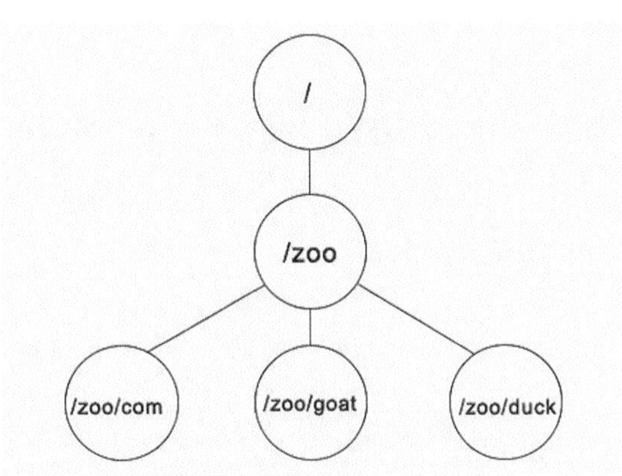

图5.3　组与组成员关系图

在 ZooKeeper 中创建组，加入组，列出组成员和删除组的示例如下。

示例 14

创建 ZooKeeper 实例和 Watcher 实例。

关键代码：

```
import org.apache.zookeeper.CreateMode;
import org.apache.zookeeper.KeeperException;
import org.apache.zookeeper.Watcher;
import org.apache.zookeeper.WatchedEvent;
```

```
import org.apache.zookeeper.ZooDefs.Ids;
import org.apache.zookeeper.ZooKeeper;
import org.junit.After;
import org.junit.Before;
import org.junit.Test;
import java.io.IOException;
import java.util.List;
/**
 * ZooKeeper 高级操作
 * 创建 ZooKeeper 实例、创建组、加入组、列出组成员、删除组
 */
public class ZookeeperAdvanced {
    public static String groupPath="/ZKGroup";
    //会话超时时间，设置为与默认时间一致
    public static final int SESSION_TIMEOUT=30000;
    //创建 ZooKeeper 实例
    ZooKeeper zk;
    //创建 Watcher 实例
    Watcher wh=new Watcher(){
        public void process(WatchedEvent event){
            System.out.println(event.toString());
        }
    };
    @Before
    public void setUp() throws Exception{
        zk = new ZooKeeper("hadoop:2181", ZookeeperAdvanced.SESSION_TIMEOUT, this.wh);
    }
    @After
    public void tearDown () throws Exception{
        zk.close();
}
```

示例 15

创建组操作。

关键代码：

```
@Test
public void CreateGroup() throws IOException, KeeperException, InterruptedException {
    //创建组
    String cGroupPath=zk.create(groupPath, "group".getBytes(), Ids.OPEN_ACL_UNSAFE,
                    CreateMode.PERSISTENT);
    //输出组路径
    System.out.println("创建的组路径为： "+cGroupPath);
}
```

示例 16

加入组操作。

关键代码：

```
@Test
public void JoinGroup() throws IOException, KeeperException, InterruptedException {
        for(int i=0;i<10;i++){
                int k=Join("/ZKGroup",i);
                //如果组不存在则退出
                if(0==k)
                    System.exit(1);
        }
}
public int Join(String groupPath,int k) throws KeeperException, InterruptedException{
        String child=k+"";
        child="child_"+child;
        //创建的路径
        String path=groupPath+"/"+child;
        //检查组是否存在
        if(zk.exists(groupPath,true) != null){
            //如果存在则加入组
            zk.create(path,child.getBytes(), Ids.OPEN_ACL_UNSAFE, CreateMode.PERSISTENT);
            return 1;
        }
        else{
            System.out.println("组不存在！");
            return 0;
        }
}
```

示例 17

列出组成员。

关键代码：

```
@Test
public void ListMembers() throws IOException, KeeperException, InterruptedException {
    //获取所有子节点
    List<String> children=zk.getChildren(groupPath, false);
    if(children.isEmpty()){
        System.out.println("组"+groupPath+"中没有组成员存在！");
        System.exit(1);
    }
    for(String child:children)
        System.out.println(child);
}
```

示例 18

删除组操作。

关键代码：

```
@Test
public void DeleteGroup() throws IOException, KeeperException, InterruptedException {
    String path = "/ZKGroup";
    try {
        List<String> children = zk.getChildren(path, false);
        for (String child : children) {
            zk.delete(path + "/" + child, -1);//-1 表示绕过版本检测机制，不管 znode 版本是什么
                                              //都直接将其删除
        }
        zk.delete(path, -1);
    } catch (KeeperException.NoNodeException e) {
        System.out.printf("Group %s does not exist\n", "/ZKGroup");
        System.exit(1);
    }
}
```

3. 使用 ZooKeeper 实现分布式系统服务器上下线的动态感知

由一个集群来提供服务，集群中的服务器会动态变化，如上线、下线或宕机等。此时要让客户端知道服务器状态的变化情况以选择一个运行良好的服务器。在服务器启动时要先向 ZooKeeper 集群注册，写入自己的信息。注册的节点是一种临时节点，最好也是序列化的节点。在客户端启动时获取服务器父目录下的子节点列表并注册监听，获取到当前在线的服务器列表并根据连接策略选择服务器连接。客户端在收到服务器状态变化的监听信息后重新获取服务器列表并注册监听。

在 ZooKeeper 中，引入了 Watcher 机制来实现分布式的通知功能。ZooKeeper 允许客户端向服务器注册一个 Watcher 监听，当服务器的一些特定事件触发了这个 Watcher，那么就会向指定客户端发送一个事件来实现分布式的通知功能。

示例 19

使用 ZooKeeper 实现分布式系统服务器上下线的动态感知。

（1）服务器端

关键步骤：

① 所有机器向 ZooKeeper 注册，注册 znode 为临时节点。

② 所有机器下线，连接断开后被 ZooKeeper 自动删除，触发监听事件。

③ 所有机器上线，触发监听事件。

关键代码：

```
import org.apache.zookeeper.*;
import org.apache.zookeeper.ZooDefs.Ids;
import org.apache.zookeeper.data.Stat;
/**
 * 服务器端代码
 */
public class DistributedServer {
```

```
    private static String connectString = "hadoop:2181";
    private static int sessionTimeout = 2000;
    private ZooKeeper zk = null;
    private String hostName;
    private String groupName = "/servers";
    public DistributedServer(String hostName) throws Exception{
        this.hostName = hostName;
        this.zk = new ZooKeeper(connectString , sessionTimeout, null);
    }
    // 将自己的信息注册到 zk 集群
    public void registToZK() throws Exception {
        // 判断父目录是否存在，不存在则创建
        Stat groupStat = zk.exists(groupName, false);
        if(groupStat == null){
            zk.create(groupName, "distributed server list".getBytes(), Ids.OPEN_ACL_UNSAFE,
                            CreateMode.PERSISTENT);
        }
        String registAddr = zk.create(groupName+"/server", hostName.getBytes(),
                        Ids.OPEN_ACL_UNSAFE, CreateMode.EPHEMERAL_SEQUENTIAL);
        System.out.println("Server is starting, reg addr：" + registAddr);
    }
    // 下线
    public void offline() throws Exception {
        zk.close();
    }
}
```

（2）客户端

关键步骤：

① 连接 ZooKeeper，获取服务器注册的 znode 并注册监听。

② 当 ZooKeeper 触发监听时，会远程调用基于接口实现的匿名内部类的 process()方法。

③ process()方法会调用 getChildren()方法获得子目录节点。

关键代码：

```
import java.util.ArrayList;
import java.util.List;
import org.apache.zookeeper.WatchedEvent;
import org.apache.zookeeper.Watcher;
import org.apache.zookeeper.ZooKeeper;
/**
 * 客户端代码
 */
public class DistributedClient {
    private static String connectString = "hadoop:2181";
```

```
        private static int sessionTimeout = 2000;
        private ZooKeeper zk = null;
        private String groupName = "/servers";
        // 开始监听服务器列表变化
        public void startListenServerListChange() throws Exception{
            this.zk = new ZooKeeper(connectString , sessionTimeout, new Watcher(){
                @Override
                public void process(WatchedEvent event) {
                    // 重新注册监听
                    try {
                        getServerList();
                    } catch (Exception e) {
                        // TODO Auto-generated catch block
                        e.printStackTrace();
                    }
                }
            });
        }
        private void getServerList() throws Exception {
            List children = zk.getChildren(groupName, true);
            List servers = new ArrayList();
            for(Object child : children) {
                byte[] data = zk.getData(groupName+"/"+child, null, null);
                servers.add(new String(data));
            }
            System.out.println("--------------当前服务器列表--------------");
            for(Object server : servers){
                System.out.println(server);
            }
            System.out.println("-----------------------------------------");
        }
    }
```

（3）服务器端测试代码

关键代码：

```
import org.junit.Test;
public class DynamicServerListTest {
    @Test
    public void testServerUp() throws Exception{
        DistributedServer server1 = new DistributedServer("server01");
        DistributedServer server2 = new DistributedServer("server02");
        server1.registToZK();
        server2.registToZK();
        Thread.sleep(10000);
        server1.offline();
```

```
            System.out.println("server01 下线...");
            Thread.sleep(10000);
            server2.offline();
            System.out.println("server02 下线...");
        }
    }
```

（4）客户端测试代码

关键代码：

```
import org.junit.Test;
public class DynamicClientTest(){
    DistributedClient client = new DistributedClient();
    client. startListenServerListChange();
    Thread.sleep(1000000);
}
```

（5）执行结果

服务器端：

```
Server is starting,reg addr:/servers/server0000000006
Server is starting,reg addr:/servers/server0000000007
server01 下线...
server02 下线...
```

客户端：

```
--------------当前服务器列表----------------
--------------------------------------------------
--------------当前服务器列表----------------
server02
server01
--------------------------------------------------
--------------当前服务器列表----------------
server02
--------------------------------------------------
--------------当前服务器列表----------------
--------------------------------------------------
```

5.3.3　技能实训

使用 ZooKeeper 实现分布式锁机制。

在程序开发过程中不得不考虑的就是并发问题。在 Java 中，对于同一个 JVM 而言，JDK 已经提供了锁和同步机制，但是在分布式环境下，往往存在多个进程对一些资源产生争用，而这些进程往往在不同的机器上，这时候 JDK 提供的功能已经不能满足实际开发的需求，就需要使用分布式锁来控制多个进程对资源的访问。

关键步骤：

（1）建立一个节点，命名为 lock。节点类型为永久节点（Persistent）。

（2）每当进程需要访问共享资源时，都会调用分布式锁的 lock()或 tryLock()方法获得锁，在第一步创建的 lock 节点下建立相应的顺序子节点，节点类型为临时顺序节点（EPHEMERAL_SEQUENTIAL），通过“name+lock+顺序号”组成特定的名字。

（3）在建立子节点后，对 lock 下面的所有以 name 开头的子节点进行排序，判断刚刚建立的子节点顺序号是否是最小的，假如是最小的子节点，则获得锁对资源进行访问。

（4）假如不是最小的子节点，就获得该节点的上一顺序节点，并给该节点注册监听事件。同时在这里阻塞，等待监听事件的发生，获得锁控制权。

（5）当调用完共享资源后，调用 unlock()方法，关闭 ZooKeeper，进而可以引发监听事件，释放锁。

本章总结

- ZooKeeper 是一个为分布式应用提供一致性服务的软件。Hadoop 生态圈中的很多技术框架都要使用 ZooKeeper 来解决单点故障问题。
- 用户可以使用 Client 命令行和 Java API 的方式管理 ZooKeeper。
- ZooKeeper 可以保证数据的强一致性，即利用这一特性可以实现分布式锁功能。

本章作业

一、简答题

1. ZooKeeper 是如何保证事务的顺序一致性的？
2. ZooKeeper 有哪几种节点类型？

二、编码题

1. 使用 ZooKeeper Cli 客户端命令创建日志服务节点 logZnode。
2. 使用 ZooKeeper Cli 客户端 set 命令对 logZnode 节点关联的字符串进行设置，将日志内容设置为“ip---时间---日志内容---日志类型”格式。
3. 使用 Java API 的方式连接 ZooKeeper 并读取 logZnode 节点内容。

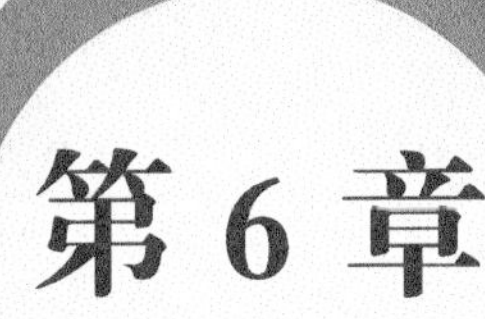

第 6 章 HBase 基础

技能目标

- 了解 HBase 体系架构
- 理解 HBase 数据模型
- 学会安装 HBase 环境

本章任务

任务 1：了解 HBase
任务 2：理解 HBase 体系架构
任务 3：理解 HBase 数据模型
任务 4：搭建 HBase 环境

本章资源下载

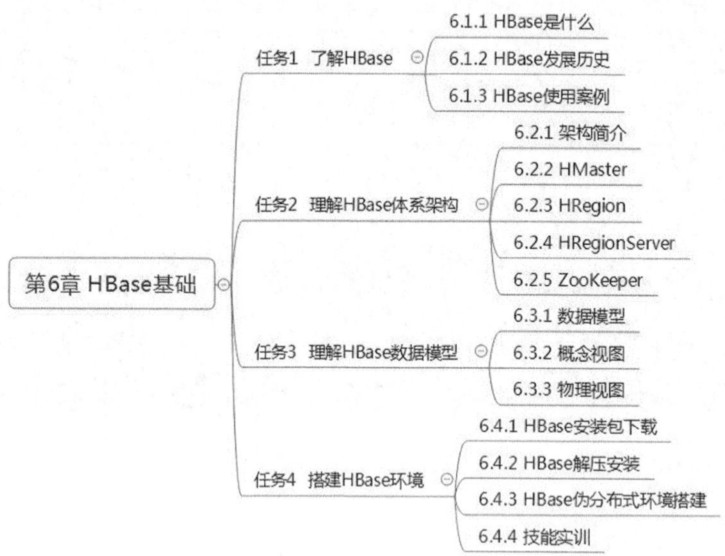

HBase 是 Apache Hadoop 中的一个子项目，它是基于 HDFS 的面向列的分布式数据库。HBase 依托于 Hadoop 的 HDFS 作为最基本存储单元，可以实现实时地随机访问超大规模结构化数据集。本章主要介绍 HBase 概念、HBase 架构设计、HBase 数据模型及如何搭建 HBase 伪分布式环境。

任务 1 了解 HBase

【任务描述】

了解 HBase 概念、发展历史以及 HBase 在企业中的使用状况。

【关键步骤】

（1）了解 HBase 概念。

（2）了解 HBase 的发展历史。

（3）了解 HBase 在企业中的应用。

6.1.1 HBase 是什么

HBase 是一个高可靠、高性能、面向列、可伸缩、可实时读写的分布式数据库，来源于 Google 发表的 Big Table 论文。就像 Big Table 利用了 Google 文件系统所提供的分布式存储一样，HBase 基于 HDFS 也提供了同 Big Table 一样的能力。HBase 基于 Hadoop 的 HDFS 作为存储单元，可以快速随机访问海量非结构化和半结构化数据。

HBase 作为可以存储海量数据的框架与 HDFS 所能处理的场景有很大区别，如表 6-1 所示。

表 6-1　HDFS 和 HBase 区别

HDFS	HBase
适合存储大量文件的分布式文件系统	建立在 HDFS 上的数据库
不支持快速单独记录查询	支持在较大的表中快速查询
支持高延迟批量处理	支持在数十亿条记录中低延迟访问单个行记录
只能顺序访问提供的数据	内部使用哈希表并提供随机接入，利用存储索引，可以在 HDFS 文件中进行快速查找

HBase 是一个分布式数据库，也不同于 Oracle、SQL Server 等关系型数据库。HBase 不支持标准 SQL 语言，也不是以行存储的关系型结构存储数据，而是以键值对的方式按列存储。HBase 是 NoSQL 数据库的一个重要代表，在 NoSQL 领域，HBase 本身并不是最优秀的，但得益于其与 Hadoop 的整合，使其在大数据领域获得了广阔的发展空间。

6.1.2　HBase 发展历史

2006 年，Google 发表了关于 Big Table 的论文。

2007 年，第一个版本的 HBase 和 Hadoop 0.15.0 一起发布。

2008 年，HBase 成为 Hadoop 的子项目。

2010 年，HBase 成为 Apache 的顶级项目。

2011 年，Cloudera 基于 HBase 0.90.1 推出 CDH3。

2012 年，HBase 发布了 0.94 版本。

2013 年，HBase 发布了 0.96 版本。

2014 年，HBase 发布了 0.98 版本。

2015～2016 年，HBase 先后发布了 1.0、1.1 和 1.2.4 版本。

2017～2018 年，HBase 先后发布了 1.3 和 1.4 版本。

6.1.3　HBase 使用案例

1．Facebook

Facebook 开发了一套 message（消息）系统，每月需要存储 1350 亿条信息，使用的存储技术就是 HBase。HBase 支持大规模数据的列级别更新，而这个特征正是消息系统（message system）所需要的。Facebook 选择 HBase 是基于以下两点原因。

（1）HBase 具有高效率的写操作、良好的随机读操作、水平扩展性、自动故障转移、强一致性等特点。

（2）HBase 基于 HDFS。

2．Alibaba

2010 年开始，HBase 成为 Alibaba 搜索系统的核心存储系统。它和计算引擎紧密结合，主要服务于搜索和推荐的业务，到目前已经有十余个版本。目前使用的版本是在社区版本的基础上经过大量优化而成。HBase 在 Alibaba 的应用场景如下。

（1）索引构建。在索引构建流程中，会将线上 MySQL 等数据库中存储的商品和用户产生的所有线上数据通过流式的方式导入到 HBase 中，并提供给搜索引擎用于构建索引。

（2）机器学习。将线上日志分析归结为商品和用户两个维度，导入分布式、持久化的消息队列，存放到 HBase 上，随线上用户的单击行为日志来产生数据更新，对应模型也随之更新，进行机器学习训练。

任务 2 理解 HBase 体系架构

【任务描述】

理解 HBase 的体系架构设计以及架构中每个组件的职责。

【关键步骤】

（1）理解 HBase 体系架构设计。

（2）理解 HBase 组件功能职责。

6.2.1 架构简介

HBase 架构原理

HBase 采用 Master/Slave 架构搭建集群。由一个 HMaster 服务器带多个 HRegionServer 服务器组成。HMaster 负责管理所有的 HRegionServer，每个 HRegionServer 负责存储许多 HRegion（HBase 逻辑表的分块）。HBase 使用 ZooKeeper 存放集群的元数据和状态信息，同时实现 HMaster 的容错处理。HBase 的架构如图 6.1 所示。

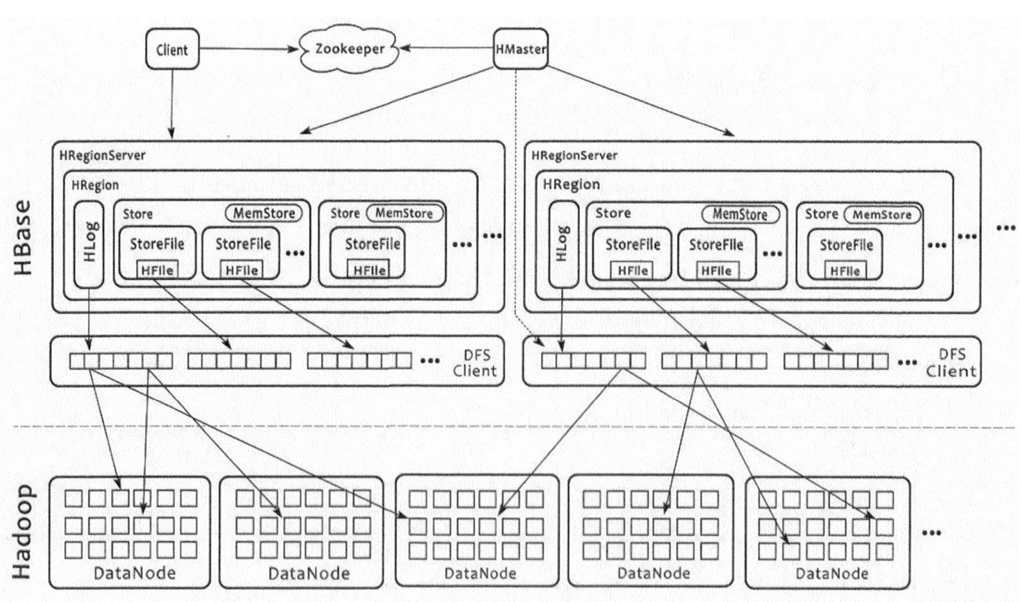

图6.1 HBase体系架构图

6.2.2 HMaster

HMaster 是 HBase 集群的主节点，主要负责 HBase 中 Table 和 Region 的管理工作，它会通知每个 HRegionServer 要维护哪些 HRegion。HMaster 并不存在单点问题，在 HBase 集群中可以启动多个 HMaster，因为 HBase 集群使用 ZooKeeper 的 Master Election 机制保证任何时刻只有一个 HMaster 在运行。一般情况下会启动两个 HMaster，非活动的 HMaster 会定期和活动 HMaster 通信以获取最新状态，从而保证自己是实时更新的。HMaster 的具体功能包括以下几点。

（1）管理 HRegionServer，通过调整 HRegion 的分布，实现负载均衡。

（2）管理和分配 HRegion。比如当 HRegion 切分的时候分配新的 HRegion 以及在 HRegionServer 停机后迁移该 HRegionServer 内的 HRegion 到其他的 HRegionServer 上等操作，都是由 HMaster 来管理的。

（3）实现 DDL（Data Definition Language，数据定义语言）操作，管理用户对表、列族以及名字空间的增、删、改、查操作。

（4）管理名字空间（Namespace）和表（Table）的元数据。

（5）管理权限控制（ACL）。

6.2.3 HRegion

HRegion 是 HBase 集群上分布式存储和负载均衡的最小单位，与 HDFS 中文件块的概念类似。HBase 使用表（Table）来组织和管理数据，当表的大小超过阈值时，HBase 会根据 RowKey（行的主键）将表水平切割成多个 HRegion。HRegion 由 HMaster 分配到相应的 HRegionServer 中，然后由 HRegionServer 负责 HRegion 的启动和管理。

HRegion 保存一个表中的一段连续的数据，每个 HRegion 都记录了它的 StartKey（开始主键）和 EndKey（结束主键），通过表名和主键范围来区分每一个 HRegion。最开始的时候，一个表只有一个 HRegion，随着数据量的增加，HRegion 逐渐变大，如果超过了设定的阈值，便将表分成两个大小基本相同的 HRegion，这个过程称为 HRegion 分裂。过程如图 6.2 所示。

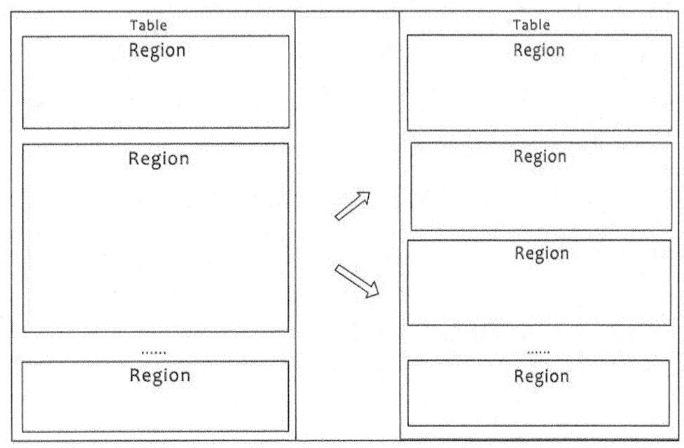

图6.2　HRegion分裂图

每个 HRegion 由多个 HStore 构成，每个 HStore 对应一个表（Table）在这个 HRegion 中的一个列族（Column Family）。HStore 是 HBase 中存储的核心，它实现了读写 HDFS 的功能，一个 HStore 由一个 MemStore 和 0 个或多个 StoreFile 组成。MemStore 是一个写缓存，用户写入数据时会先放入 MemStore，再由 MemStore 根据一定的算法将数据刷入（Flush）到 StoreFile 中。StoreFile 是 HBase 中的最小存储单元，底层由 HFile 实现，用于存储 HBase 的数据，实质上是 HDFS 的二进制格式文件。

6.2.4 HRegionServer

HRegionServer 一般和 DataNode 在同一台机器上，以实现数据的本地性。一般情况下，一台机器上只运行一个 HRegionServer。HRegionServer 负责数据的读写服务，用户通过 HRegionServer 可以实现对数据的访问，同时 HRegionServer 还管理本地 Region 和 Table 中的数据。HRegion 由两部分组成：WAL（早期版本称为 HLog）部分和 HRegion 部分。HRegion 前面已经介绍过了，WAL 即 Write Ahead Log，它是 HDFS 上的一个文件，所有数据的写操作都会先保证写入到这个 Log 文件后，才能真正更新 MemStore，最后写入到 HFile 中，这种模式可以保证即使 HRegionServer 宕机，依然可以从 Log 文件中读取数据，达到故障恢复的效果。

6.2.5 ZooKeeper

ZooKeeper 在 HBase 中起到非常重要的作用，具体功能如下。

（1）HBase 使用 ZooKeerer 维护集群中服务器的状态并协调分布式系统的工作。

（2）ZooKeeper 维护服务器是否存活、是否可访问的状态，并提供服务器故障/宕机的通知。

（3）ZooKeeper 使用一致性算法来保证服务器之间的同步。

（4）ZooKeeper 负责 HMaster 的选举工作。

（5）RootRegion 管理。

ZooKeeper 存储了 HBase 中两张特殊的数据表的位置，分别是-ROOT-（根数据表）和.META.（元数据表）。.META.表记录普通表的 HRegion 标识符信息，随着 HRegion 的分裂，.META.表信息会不断增加，增加到一定阈值时，.META.表会被分割为多个 HRegion。HBase 中使用-ROOT-表来保存.META.的 HRegion 信息，而-ROOT-表示不能被分割，也就是说-ROOT-表只能有一个 HRegion。客户端在访问用户数据前，需要先访问 ZooKeeper，获取-ROOT-表中的.META.信息，接着访问.META.表，最后才能获取到用户数据所在的位置进行访问。

任务 3 理解 HBase 数据模型

【任务描述】

理解 HBase 数据模型中涉及的术语。

【关键步骤】

（1）掌握 HBase 数据模型。

（2）理解 HBase 物理视图。

（3）理解 HBase 概念视图。

6.3.1 数据模型

HBase 的数据模型是由一张张的表组成的，表由行和列组成。但是 HBase 数据库中的行和列又和关系型数据库中的稍有不同。下面介绍 HBase 数据模型中一些术语的概念。

（1）表（Table）：HBase 会将数据组织成一张张的表，表是稀疏表（NULL 数据不存储），表的索引是行关键字、列关键字和时间戳。需要注意的是，表名必须是能用在文件路径里的合法名字，因为 HBase 的表将映射成 HDFS 中的文件。

（2）行（Row）：表中的每一行代表一个数据对象，每一行都以一个行关键字（Row Key）进行唯一标识。

（3）行关键字（Row Key）：行的主键，唯一标识一行数据，也被称为行键。表中的行根据 Row Key 的字典序进行排序。设计 Row Key 的时候要充分利用排序存储的特性，将经常一起读取的行存储到一起。行键在添加数据时首次被确定，所有对表的数据操作都必须要通过表的行键。

（4）列族（Column Family）：HBase 表中的每一列都归属于某个列族，列族是表的 Schema 的一部分，而列不是。在定义 HBase 表的时候需要提前设置好列族，列族一旦确定后，就不能轻易修改。列族名称不能包含 ASCII 控制字符（ASCII 码在 0～31，外加 127）和冒号（:）。

（5）列关键字（Column Qualifier）：也称列键。列族中的数据通过列键来进行映射，格式为“<family>:<qualifier>”，其中，family 为列族名，qualifier 为列族修饰符。列键不是表的 Schema 的一部分，所以列族修饰符及其对应的值可以动态增加或删除。

（6）存储单元格（Cell）：每一个行键、列关键字共同组成一个单元格，在该单元格中存储数据。

（7）时间戳（Timestamp）：在向 HBase 表中插入数据时都会使用时间戳来进行版本标识，作为单元格数据的版本号。每一个列族的单元格数据的版本数量都由 HBase 单独维护，默认情况下 HBase 保留三个版本数据。

6.3.2 概念视图

HBase 中的表可以看成是一个大的映射关系，通过行键、时间戳、列关键字（family:qualifier）可以定位到单元格中的数据。在 HBase 中，表格中的单元格如果是空，将不占用空间。表的概念视图如图 6.3 所示。

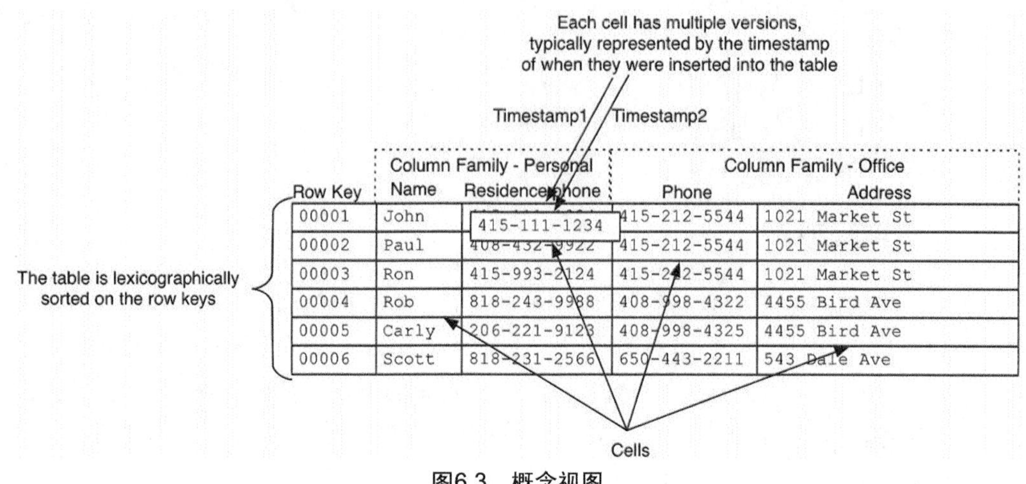

图6.3 概念视图

从图 6.3 中可以看出，表由两个列族组成，分别是 Personal 和 Office，每个列族包含两个列，包含数据的实体称为单元格。每一个单元格可以有多个版本，通过插入数据时间戳来表示不同版本。行数据会基于 Row Key 进行排序。

为了方便理解，可以将 HBase 数据模型想象成多维映射，图 6.3 中第一行数据的多维映射如图 6.4 所示。

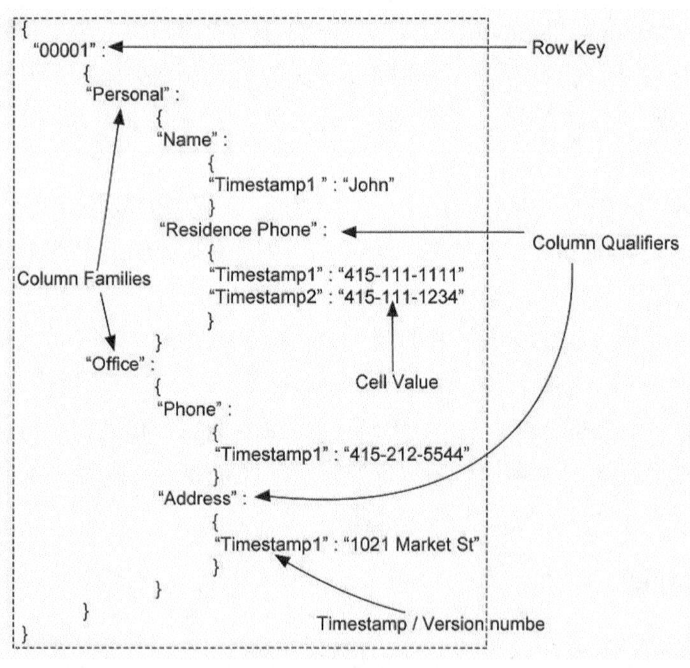

图6.4 概念视图——映射

从图 6.4 中可以看出，Row Key 映射到 Column Family 的列表，每一个 Column Family 映射到 Column Qualifier 的列表，每一个 Column Qualifier 映射到 Timestamp 的列表，每一个 Timestamp 映射到一个值（Cell 的值）。

6.3.3　物理视图

尽管在 HBase 的概念视图中，表格被视为一组稀疏的行的集合，但在物理存储上它们是按列族进行存储的，可以随时将新的列族修饰符添加到现有的列族。图 6.3 中第一行数据对应的物理视图如表 6-2～表 6-5 所示。

表 6-2　列 Personal:name

<table>
<tr><th rowspan="2">Row Key</th><th rowspan="2">Timestamp</th><th colspan="2">Family:Qaulifier</th></tr>
<tr><th>列</th><th>值</th></tr>
<tr><td>00001</td><td>Timestamp1</td><td>Personal:Name</td><td>John</td></tr>
</table>

表 6-3　列 Personal:Residencephone

<table>
<tr><th rowspan="2">Row Key</th><th rowspan="2">Timestamp</th><th colspan="2">Family:Qaulifier</th></tr>
<tr><th>列</th><th>值</th></tr>
<tr><td rowspan="2">00001</td><td>Timestamp1</td><td rowspan="2">Personal:Residencephone</td><td>415-111-1111</td></tr>
<tr><td>Timestamp2</td><td>415-111-1234</td></tr>
</table>

表 6-4　列 Office:Phone

<table>
<tr><th rowspan="2">Row Key</th><th rowspan="2">Timestamp</th><th colspan="2">Family:Qaulifier</th></tr>
<tr><th>列</th><th>值</th></tr>
<tr><td>00001</td><td>Timestamp1</td><td>Office:Phone</td><td>415-212-5544</td></tr>
</table>

表 6-5　列 Office:Address

<table>
<tr><th rowspan="2">Row Key</th><th rowspan="2">Timestamp</th><th colspan="2">Family:Qaulifier</th></tr>
<tr><th>列</th><th>值</th></tr>
<tr><td>00001</td><td>Timestamp1</td><td>Office:Address</td><td>1021 Market St</td></tr>
</table>

HBase 就是这样一种基于列模式的映射数据库，只能表示简单的“键->值”的映射关系，与关系型数据库相比，它具有以下特点。

（1）数据类型。HBase 中只有字符串类型，也就是说，HBase 只保存字符串；而在关系型数据库中提供了更多的数据类型可供选择存储。

（2）数据操作。HBase 只提供数据插入、查询、删除、清空等操作，表和表之间不存在关联操作；而关系型数据库中会有很多表间的关联操作。

（3）存储模式。这是 HBase 与关系型数据库最明显的区别，HBase 是基于列存储的，每个列族由几个文件保存，不同列族的文件是分离的；而关系型数据库是基于表格结构和行模式存储的。

（4）数据更新。HBase 的更新操作实际上是插入了新的数据，旧版本依然会根据时间戳进行不同版本的保存；而关系型数据库的更新是对数据的替换修改。

（5）可扩展性。HBase 的性能提高可以通过简单地增加机器的方式实现，而 HBase 本身就是为了这个目的开发出来的。HBase 具有很好的容错机制，而关系型数据库需要增加中间层才能实现类似的功能。

需要注意的是，在概念视图上有些列是空白的，这样的列实际上并不会被存储，当请求这些空白的单元格时，会返回 NULL 值。如果在查询的时候不提供时间戳，会返回离现在最近的版本的数据，因为在不同版本存储的时候，数据会按照时间戳来排序。

任务 4 搭建 HBase 环境

【任务描述】

学会搭建 HBase 伪分布式环境。

【关键步骤】

（1）下载 HBase 安装包。

（2）解压安装。

（3）配置并启动 HBase。

6.4.1 HBase 安装包下载

HBase 的安装有三种方式，分别为单机模式、伪分布模式、完全分布模式。

（1）单机模式比较简单，只需要在一个节点上安装配置 HBase，此模式下只需在 HBase 配置文件 hbase-site.xml 中指定 HBase 的文件存储目录即可，配置方式如下。

```
<configuration>
    <property>
        <name>hbase.rootdir</name>
        <value>file:///home/hadoop/ hbase</value>
    </property>
</configuration>
```

其中，hbase.rootdir 参数指定了 HBase 的文件存储目录，上面的配置中使用的是 Linux 系统的文件目录，也可以替换为 HDFS 的目录。

（2）伪分布模式的 HBase 也是在单个节点上运行，和单机模式一样，但在该节点上会同时运行 HMaster、HRegionServer 及 HQuorumPeer（如果采用 HBase 内置的 ZooKeeper）三个进程。

（3）完全分布模式与伪分布模式的主要区别是完全分布模式的 HBase 运行在多个机器上，通常是将 HBase 的 HMaster 运行在 HDFS 的 NameNode 上，而将 HRegionServer 运行在 HDFS 的 DataNode 上。

本节主要介绍 HBase 的伪分布模式安装，采用的版本是 CDH 的 HBase-1.2.0-cdh5.14.2，读者可以到 Cloudera 官网去下载对应的版本，将软件下载到合适的位置，本书选择放在安装用户“hadoop”home 目录下的 software 目录下。安装 HBase 伪分布式环境的详细步骤请扫描二维码。

HBase 伪分布式环境搭建

安装 HBase 之前，需要具备如下三个前置条件。

（1）HBase 的运行需要 JDK，安装 HBase 前需要先自行安装 JDK。

（2）本章前面介绍了 HBase 底层存储是依赖于 HDFS 的，因此安装 HBase 的前提是必须安装 Hadoop 环境。关于 Hadoop 环境的安装可以参考第 1 章内容。

（3）HBase 依赖于 ZooKeeper 来做分布式协调工作，所以需要先有 ZooKeeper 的环境。目前版本中，HBase 内置了 ZooKeeper 环境，也可以使用外部自己搭建的 ZooKeeper 环境。本节是使用自己搭建的 ZooKeeper 环境。关于 ZooKeeper 环境的搭建可以参考第 5 章内容。

6.4.2 HBase 解压安装

软件包下载完成以及三个前置条件都准备好以后，就需要将安装包解压。命令如下。

```
[hadoop@hadoop ~]$ tar –zxvf /home/hadoop/software/hbase-1.2.0-cdh5.14.2.tar.gz
```

将解压后的文件复制到/opt 目录下。

```
[hadoop@hadoop ~]$ sudo mv ~/software/hbase-1.2.0-cdh5.14.2 /opt
```

切换到 HBase 解压目录，可以查看 HBase 的目录结构。

```
[hadoop@hadoop ~]$ cd /opt/ hbase-1.2.0-cdh5.14.2
```

HBase 常用目录结构及描述如下。

- bin：存放 HBase 所有可执行命令与脚本。
- conf：HBase 配置文件存放目录。
- hbase-webapps：存放 Web 应用的目录，这些应用可以查看 HBase 的运行状态。默认的访问地址是 http://hadoop:16010，其中，hadoop 是 HBase Master 节点的机器名。
- lib：存放 HBase jar 文件，包括第三方依赖包以及与 Hadoop 相关的 jar 文件。其中与 Hadoop 相关的 jar 包最好能与实际运行的 Hadoop 版本一致，以保证稳定运行。
- logs：日志存放目录。

解压完成后，将 HBase 的安装目录添加到环境变量中，命令如下。

```
[hadoop@hadoop ~]$ vi ~/.bashrc
export HBASE_HOME=/opt/hbase-1.2.0-cdh5.14.2
export PATH=.:$HBASE_HOME/bin:$PATH
```

注意

JDK、Hadoop、ZooKeeper 的安装目录也需要配置到环境变量中。

修改完成以后，还需要让环境变量生效。

```
[hadoop@hadoop ~]$ source ~/.bashrc
```

6.4.3 HBase 伪分布式环境搭建

1. 配置

HBase 安装包解压完成并配置好环境变量以后，就需要对 HBase 进行配置。HBase 的配置文件存放目录是在安装目录的 conf 目录下。伪分布式环境搭建需要配置两个文件：hbase-env.sh 文件和 hbase-site.xml 文件。

（1）配置 hbase-env.sh 文件

hbase-env.sh 文件用来配置全局 HBase 集群系统的特性。在这个文件中，需要添加 JDK 的安装路径，同时还要配置 ZooKeeper。配置方式如下。

切换到 HBase 安装目录的 conf 目录下。

```
[hadoop@hadoop ~]$ cd /opt/ hbase-1.2.0-cdh5.14.2/conf
```

编辑文件并添加配置内容。

示例 1

配置 HBase 的 hbase-env.sh 文件。

```
[hadoop@hadoop conf]$ vi hbase-env.sh
export JAVA_HOME=/opt/jdk1.8.0_171
export HBASE_CLASSPATH=/opt/hadoop-2.6.0-cdh5.14.2/etc/hadoop
export HBASE_MANAGES_ZK=false
```

配置说明如下。

- export JAVA_HOME=/opt/jdk1.8.0_171：配置 JDK 的安装目录。
- export HBASE_CLASSPATH=/opt/hadoop-2.6.0-cdh5.14.2/etc/hadoop：配置 Hadoop Master 配置文件的路径，也就是 core-site.xml 的路径。
- export HBASE_MANAGES_ZK=false：该属性可以设置 true 和 false 两个值，默认是 true，表示使用 HBase 自带的 ZooKeeper，false 表示使用外部安装的 ZooKeeper。由于本节使用自己安装的 ZooKeeper，所以设置为 false。

（2）配置 hbase-site.xml 文件

hbase-site.xml 文件同样是放在 HBase 安装目录的 conf 目录下，它的配置如下。

切换到 HBase 安装目录的 conf 目录下。

```
[hadoop@hadoop ~]$ cd /opt/ hbase-1.2.0-cdh5.14.2/conf
```

编辑文件并添加配置内容。

示例 2

配置 HBase 的 hbase-site.xml 文件。

```
[hadoop@hadoop conf]$ vi hbase-site.xml
<configuration>
    <property>
        <name>hbase.rootdir</name>
        <value>hdfs://hadoop:8020/hbase</value>
    </property>
    <property>
        <name>hbase.master</name>
        <value>hadoop:60000</value>
    </property>
    <property>
        <name>hbase.cluster.distributed</name>
        <value>true</value>
```

```
    </property>
    <property>
        <name>hbase.zookeeper.quorum</name>
        <value>hadoop</value>
    </property>
    <property>
        <name>hbase.zookeeper.property.clientPort</name>
        <value>2181</value>
    </property>
</configuration>
```

配置说明如下。

- hbase.rootdir：配置 HBase 数据存放目录，这里使用的是 HDFS，注意这个值需要与安装的 Hadoop 目录下 etc/hadoop/core-site.xml 文件中的 fs.defaultFS 属性值相对应。例如 defaultFS 的值设置为 hdfs://hadoop:8020/，那么 hbase.rootdir 的值就需要设置为 hdfs://hadoop:8020/hbase。
- hbase.master：指定 HBase 的 HMaster 服务器的地址、端口。这里配置 HMaster 服务器为 hadoop，端口为 60000（默认端口）。
- hbase.cluster.distributed：默认为 false，表示单机运行，如果设置为 true，表示在分布模式下运行。由于本节采用的是伪分布模式，所以配置为 true。
- hbase.zookeeper.quorum：该属性配置的是 ZooKeeper 集群各服务器位置，一般为奇数个服务器。本节采用的是外部伪分布模式安装的 ZooKeeper，所以配置的值是 hadoop。如果有多个 ZooKeeper 节点，配置时要使用逗号进行分隔。
- hbase.zookeeper.property.clientPort：该属性配置的是 ZooKeeper 的端口号，配置自己搭建的 ZooKeeper 集群时，这个属性需要与 ZooKeeper 安装目录下 conf/zoo.cfg 文件中的 clientPort 属性值一致。

2. 启动

配置完成以后，就可以启动 HBase。启动和停止等脚本命令都存放在 HBase 安装目录的 bin 目录下。在 HBase 启动之前，需要先启动 Hadoop 以及 ZooKeeper，启动顺序为先启动 HDFS，再启动 ZooKeeper，最后启动 HBase；关闭的时候相反，需要先关闭 HBase，再关闭 ZooKeeper，最后关闭 HDFS。HBase 启动和停止命令如下。

启动命令：$HBASE_HOME/bin/ start-hbase.sh

停止命令：$HBASE_HOME/bin/ stop-hbase.sh

示例 3

启动配置完成的 HBase。

```
[hadoop@hadoop ~]$ cd /opt/hbase-1.2.0-cdh5.14.2/bin
[hadoop@hadoop bin]$ sh start-hbase.sh
```

3. 验证

HBase 启动完成以后，通常可以使用以下几种方式进行验证。

（1）通过 jps 命令查看启动的进程。

```
[hadoop@hadoop ~]$ jps
16051 ResourceManager
15622 NameNode
16152 NodeManager
17000 HRegionServer
15723 DataNode
17404 Jps
15903 SecondaryNameNode
16511 QuorumPeerMain
16863 HMaster
```

命令执行完以后，可以看到启动的进程包括 Hadoop 中的 NameNode、DataNode、SecondaryNameNode、ResourceManager、NodeManager 进程，以及 ZooKeeper 的 QuorumPeerMain 进程，同时多了 HMaster、HRegionServer 两个进程，这就说明 HBase 已经启动成功。

（2）通过 webui 的方式验证。

HBase 启动后，可以通过 webui 的方式查看运行环境。网址为：http://hadoop:60010，其中，hadoop 为 HBase 安装的 HMaster 节点的机器名，60010 为默认端口号。

（3）查看 Hadoop 及 ZooKeeper 是否有/hbase 目录和节点。

示例 4

查看 HDFS 文件系统中是否有/hbase 目录。

```
[hadoop@hadoop ~] hdfs dfs -ls /
drwxr-xr-x    - hadoop supergroup          0 2018-07-09 16:42 /hbase
```

从上面的结果中可以看出，启动 HBase 以后，会在 HDFS 文件系统中自动创建 HBase 数据存放目录/hbase，原因是前面在 hbase-site.xml 中配置 hbase.rootdir 属性值为 hdfs://hadoop:8020/hbase。

示例 5

查看 ZooKeeper 中是否有 hbase 节点。

使用客户端连接 ZooKeeper：

```
[hadoop@hadoop ~] zkCli.sh -server 127.0.0.1:2181
[zk: 127.0.0.1:2181(CONNECTED) 0]
```

使用 ls 命令查看根节点下的内容：

```
[zk: 127.0.0.1:2181(CONNECTED) 0] ls /
[zookeeper, hbase]
```

从上面的结果中可以看出，HBase 启动成功之后会在 ZooKeeper 根节点下自动创建 hbase 节点。

从示例 4 和示例 5 可以看出，通过查看 HDFS 和 ZooKeeper 的目录和节点，也可以验证 HBase 是否启动成功。

（4）使用 Hbase Shell 方式验证。

HBase 还提供了 Shell 的方式来操作，启动 HBase 后，在命令行输入“hbase shell”，可以进入 HBase Shell 命令行。

示例 6

进入 HBase Shell 命令行。

```
[hadoop@hadoop ~]$ cd /opt/hbase-1.2.0-cdh5.14.2/bin
[hadoop@hadoop bin]$ hbase shell
```

进入到 Shell 环境以后，就可以使用 HBase Shell 命令进行操作。

示例 7

使用 create 命令创建 Student（学生）表，包含两个列族“info”和“course”。

```
hbase(main):000:0> create 'Student',{NAME => 'info'}, {NAME => 'course'}
0 row(s) in 23.1190 seconds

=> Hbase::Table - Student
```

表创建完成后，可以查看 HBase 中是否存在刚创建的表。

示例 8

使用 list 命令查看 Student 表是否存在于数据库中。

```
hbase(main):001:0> list
TABLE
Student
```

从结果可以看出，示例 7 创建的表 Student 已经存在，证明创建成功。也就是说，HBase 的环境已经搭建成功。HBase Shell 的其他命令会在第 7 章详细讲解，这里读者只需会使用命令进行验证即可。

6.4.4 技能实训

搭建自己的 HBase 伪分布式环境。

实现步骤：

（1）下载安装包，解压安装并配置环境变量（注意：需要先成功安装 JDK、Hadoop 及 ZooKeeper）。

（2）对 HBase 的配置文件进行配置，主要是配置 hbase-env.sh 和 hbase-site.xml 两个文件。

（3）启动 HBase，启动的命令为 start-hbase.sh，在$HBASE_HOME/bin 目录下。

（4）验证 HBase 是否启动成功，可以通过 jps 命令查看进程、webui、查看 HDFS 目录和 ZooKeeper 节点、启动 HBase Shell 的方式进行验证。

本章总结

- HBase 是一个分布式的面向列的数据库，能够实时地随机访问超大规模数据集，是对 HDFS 的有力补充。
- HBase 的体系架构包括 HMaster、HRegionServer、HRegion、ZooKeeper。
- HBase 中的数据会按照 Row Key 进行字段排序，Row Key 设计得好，将对 HBase 的性能有极大的提升。

➢ HBase 使用 ZooKeeper 做分布式协调工作，所以 HMaster 不存在单点故障问题。

本章作业

简答题

1. 简述 HBase 写数据。
2. HBase 的数据模型有哪些？
3. HBase 的 Row Key 设计原则有哪些？
4. 简述 HBase 的特点有哪些。
5. HBase 的分布式存储的最小单元是什么？

第7章

HBase 操作

技能目标

- 熟练使用 HBase Shell 操作 HBase
- 熟练使用 HBase Java API 操作 HBase
- 会使用 Rest 方式操作 HBase

本章任务

任务 1　使用 HBase Shell 完成《王者荣耀》游戏玩家信息管理操作

任务 2　使用 HBase Java API 完成《王者荣耀》游戏玩家信息管理操作

任务 3　使用 HBase Rest API 访问《王者荣耀》游戏玩家信息表

本章资源下载

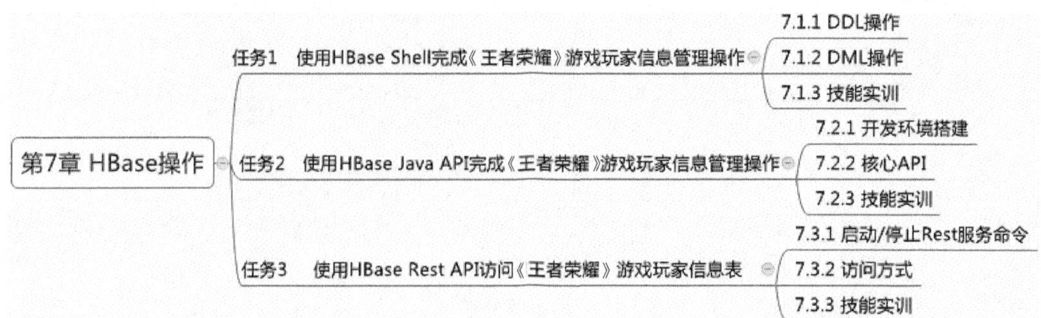

HBase 作为一个分布式面向列的数据库，提供了多种操作数据库的方式，包括 HBase Shell 方式、Java API 方式、Rest API 方式等。本章主要讲解使用 HBase Shell、Java API 以及 Rest API 的方式操作 HBase。

任务 1 使用 HBase Shell 完成《王者荣耀》游戏玩家信息管理操作

【任务描述】

本任务主要使用 HBase Shell 的方式实现对《王者荣耀》游戏玩家信息表的操作，包括表的创建、删除及数据的插入、查询等操作。

【关键步骤】

（1）创建表。

（2）修改表。

（3）插入数据。

（4）查询表中数据。

（5）删除表。

7.1.1 DDL 操作

HBase 提供了 Shell 的方式来操作和管理表，在前面一章测试 HBase 启动的时候已经简单地使用了 HBase Shell 的 list 和 create 命令，本节主要介绍使用 HBase Shell 的方式来完成 HBase 数据库的 DDL（Data Definition Language，数据定义语言）操作。

DDL 主要是对表的一些管理操作，HBase Shell 中的常用 DDL 命令如表 7-1 所示。

表 7-1 HBase Shell 常用 DDL 命令

HBase Shell 命令	功能描述
list	列出 HBase 中的所有表
create	创建一张表
describe	列出表的详细信息
alter	修改表的列族

续表

HBase Shell 命令	功能描述
disable	禁用表，使表无效
enable	启用表，使表有效
drop	删除一张表
exists	判断表是否存在

下面使用 HBase Shell 的 DDL 操作来完成《王者荣耀》游戏玩家信息表的建立、修改及删除操作。

1. 创建《王者荣耀》游戏玩家信息表并查看表详细信息

（1）列出 HBase 中所有的表

【命令】

```
list
```

示例 1

查看已安装的 HBase 环境中的表。

输入命令：

```
hbase(main):000:0> list
```

输出结果：

```
TABLE
0 row(s) in 0.0270 seconds

=> []
```

从结果中可以看出，由于是新安装的 HBase 环境，所以数据库中目前是没有表存在的。

（2）创建一张表

【命令】

```
create 'table', {NAME => 'family', VERSIONS => <VERSIONS>}
```

其中，表名称 table，列族名称 family 必须用单引号括起来。如果有多个列族，应以逗号分隔，每个列族单独用{}括起来。指定列族参数的格式为参数名=>参数值，注意赋值符号为“=>”，参数名必须大写。

示例 2

创建《王者荣耀》游戏玩家信息表 gamer，包含列族 personalInfo（个人信息）、recordInfo（战绩信息）以及 assetsInfo（资产信息）。

输入命令：

```
hbase(main):001:0> create 'gamer',{NAME => 'personalInfo', VERSIONS => 1},{NAME => 'recordInfo', VERSIONS => 1,{NAME => 'assetsInfo', VERSIONS => 1}
```

输出结果：

```
0 row(s) in 5.2800 seconds
=> Hbase::Table – gamer
```

gamer 表创建成功后，可以使用 list 命令查看 HBase 中是否有 gamer 表。

输入命令：

```
hbase(main):002:0> list
```

输出结果：

```
TABLE
gamer
1 row(s) in 0.0190 seconds

=> [ "gamer"]
```

从结果中可以看出，gamer 表已经在 HBase 数据库中，证明表已经创建成功。

在创建表时，除了列族名称，列族的其余参数均为可选项。示例 2 中创建 gamer 表的代码可以简化为下面的形式。

```
create 'gamer', 'personalInfo', 'recordInfo', 'assetsInfo'
```

（3）查看表的详细信息

【命令】

describe '表名称'

其中，“表名称”必须用单引号括起来。

示例 3

查看 gamer 表的详细信息。

输入命令：

```
hbase(main):003:0> describe 'gamer'
```

输出结果

```
Table gamer is ENABLED
gamer
COLUMN FAMILIES DESCRIPTION
{NAME => 'assetsInfo', BLOOMFILTER => 'ROW', VERSIONS => '1', IN_MEMORY => 'false', KEEP_DELETED_CELLS => 'FALSE', DATA_BLOCK_ENCODING => 'NONE', TTL => 'FOREVER', COMPRESSION => 'NONE', MIN_VERSIONS => '0', BLOCKCACHE => 'true', BLOCKSIZE =>'65536', REPLICATION_SCOPE => '0'}
{NAME => 'personalInfo', BLOOMFILTER => 'ROW', VERSIONS => '1', IN_MEMORY => 'false', KEEP_DELETED_CELLS => 'FALSE', DATA_BLOCK_ENCODING => 'NONE', TTL => 'FOREVER', COMPRESSION => 'NONE', MIN_VERSIONS => '0', BLOCKCACHE => 'true', BLOCKSIZE=> '65536', REPLICATION_SCOPE => '0'}
{NAME => 'recordInfo', BLOOMFILTER => 'ROW', VERSIONS => '1', IN_MEMORY => 'false', KEEP_DELETED_CELLS => 'FALSE', DATA_BLOCK_ENCODING => 'NONE', TTL => 'FOREVER', COMPRESSION => 'NONE', MIN_VERSIONS => '0', BLOCKCACHE => 'true', BLOCKSIZE =>'65536', REPLICATION_SCOPE => '0'}
3 row(s) in 0.3910 seconds
```

从结果中可以看出，使用 describe 命令可以查看表中的列族信息，列族描述信息的含义如表 7-2 所示。

表 7-2　列族描述信息含义

列族描述	可选值	含义
NAME	可打印的字符串	列族名称，参考 ASCII 码表中可打印字符
BLOOMFILTER	NONE（默认）|ROWCOL|ROW	提高随机读的性能
VERSION	数字	列族中单元时间版本最大数量
IN_MEMORY	true|false，默认 false	使得列族在缓存中拥有更高优先级
KEEP_DELETED_CELLS	TRUE|FALSE（默认）	启用后可以避免被标记为删除的单元从 HBase 中删除
DATA_BLOCK_ENCODING	NONE（默认）	数据库编码
TTL	默认 FOREVER	单元时间版本超时时间，可指定多长时间（秒）后失效
COMPRESSION	NONE（默认）|LZO|SNAPPY|GZIP	压缩算法
MIN_VERSIONS	数字	列族中单元时间版本最小数量
BLOCKCACHE	true|false，默认 true	是否将数据放入读缓存
BLOCKSIZE	默认 65536 字节	数据块大小，数据块越小，索引越大
REPLICATION_SCOPE	默认 0	开启复制功能

2. 修改表的列族信息

使用 alter 命令可以为表增加或修改列族。

【命令】

```
alter 'table', {NAME => 'family', VERSIONS => <VERSIONS>}
```

其中，列族名称参数 NAME 必须提供，如果已经存在，则修改，否则会增加一个列族。

示例 4

将示例 2 中创建的 gamer 表中的 personalInfo 列族的 TTL 信息设置为 180 天（15552000 秒）。

注意

TTL（Time to Live）用于限定数据的超时时间。

输入命令：

```
hbase(main):004:0> alter 'gamer',{NAME=>'recordInfo',TTL=>'15552000'}
```

输出结果：

```
Updating all regions with the new schema...
0/1 regions updated.
1/1 regions updated.
Done.
0 row(s) in 27.5400 seconds
```

修改完以后，使用 describe 命令查看表的信息。

输入命令：

```
hbase(main):005:0> describe 'gamer'
```

输出结果：

```
Table gamer is ENABLED
gamer
COLUMN FAMILIES DESCRIPTION
{NAME => 'assetsInfo', BLOOMFILTER => 'ROW', VERSIONS => '1', IN_MEMORY => 'false', KEEP_DELETED_CELLS => 'FALSE', DATA_BLOCK_ENCODING => 'NONE', TTL => 'FOREVER', COMPRESSION => 'NONE', MIN_VERSIONS => '0', BLOCKCACHE => 'true', BLOCKSIZE => '65536', REPLICATION_SCOPE => '0'}
{NAME => 'personalInfo', BLOOMFILTER => 'ROW', VERSIONS => '1', IN_MEMORY => 'false', KEEP_DELETED_CELLS => 'FALSE', DATA_BLOCK_ENCODING => 'NONE', TTL => 'FOREVER', COMPRESSION => 'NONE', MIN_VERSIONS => '0', BLOCKCACHE => 'true', BLOCKSIZE => '65536', REPLICATION_SCOPE => '0'}
{NAME => 'recordInfo', BLOOMFILTER => 'ROW', VERSIONS => '1', IN_MEMORY => 'false', KEEP_DELETED_CELLS => 'FALSE', DATA_BLOCK_ENCODING => 'NONE', TTL => '15552000 SECONDS (180 DAYS)', COMPRESSION => 'NONE', MIN_VERSIONS => '0', BLOCKCACHE => 'true', BLOCKSIZE => '65536', REPLICATION_SCOPE => '0'}
3 row(s) in 0.2920 seconds
```

从结果中可以看出，gamer 表中 recordInfo 的 TTL 信息已经修改成'15552000 SECONDS (180 DAYS)'。

3. **删除表**

使用 drop 命令可以对表进行删除操作。在示例 3 中，使用 describe 命令查看表详细信息的时候，输出结果的第一行“Table gamer is ENABLED”中，ENABLED 是 HBase 的一种状态。在 HBASE 中，表分为两种状态：ENABLED 和 DISABLED，分别表示表是否可用。当表为 ENABLED 状态时会被禁止删除，需要将表设置为 DISABLED 状态才可以删除。

（1）disable 命令

使用 disable 命令可以禁用表。

【命令】

```
disable '表名'
```

示例 5

将示例 2 创建的 gamer 表禁用。

输入命令：

```
hbase(main):006:0> disable 'gamer'
```

输出结果：

```
0 row(s) in 2.4040 seconds
```

禁用完以后，可以使用 is_disabled 命令查看表是否被禁用。

输入命令：

```
hbase(main):007:0> is_disabled 'gamer'
```

输出结果：

```
true
0 row(s) in 0.0530 seconds
```

从结果中的“true”标识可以看出，gamer 表已经被禁用。

（2）enable 命令

使用 enable 命令可以重新启用表。

【命令】

```
enable '表名'
```

示例 6

将示例 5 禁用的 gamer 表重新启用。

输入命令：

```
hbase(main):008:0> enable 'gamer'
```

输出结果：

```
0 row(s) in 1.4040 seconds
```

启用完以后，可以使用 is_enabled 命令查看表是否被启用。

输入命令：

```
hbase(main):009:0> is_enabled 'gamer'
```

输出结果：

```
true
0 row(s) in 0.0530 seconds
```

从结果中的“true”标识可以看出，gamer 表已经被启用。

（3）drop 命令

使用 drop 命令可以删除表。

【命令】

```
drop '表名称'
```

示例 7

删除示例 2 创建的 gamer 表。

先设置 gamer 表的状态为 DISABLED。

输入命令：

```
hbase(main):010:0> disable 'gamer'
```

输出结果：

```
0 row(s) in 2.4040 seconds
```

表禁用之后，就可以删除了。

输入命令：

```
hbase(main):011:0> drop 'gamer'
```

输出结果：

```
0 row(s) in 1.5300 seconds
```

（4）exists 命令

使用 exists 命令可以查看表是否存在。

【命令】

exists '表名称'

示例 8

查看示例 7 删除的 gamer 表是否存在。

输入命令：

```
hbase(main):012:0> exists 'gamer'
```

输出结果：

```
Table gamer does not exist
0 row(s) in 0.0530 seconds
```

从结果中可以看出，表 gamer 已经不存在。

7.1.2 DML 操作

7.1.1 节介绍了 HBase Shell 中的 DDL 操作命令，本节主要介绍使用 DML 操作命令来完成对 HBase 表数据的操作，包括添加数据、查询数据、删除数据等。由于在示例 7 中演示删除表操作，把创建的《王者荣耀》游戏信息表 gamer 删除了，需要先把表重新创建出来，请参考示例 2 完成。

DML 主要是对表数据的操作，HBase Shell 中的常用 DML 命令如表 7-3 所示。

表 7-3 HBase Shell 常用 DML 命令

HBase Shell 命令	功能描述
put	向指定的单元添加值
scan	通过扫描表来获取数据
get	获取行或者单元的值
count	统计表中行的数量，一个行键为一行
delete	删除指定对象的值
deleteall	删除整行
truncate	清空表的数据

下面就使用 HBase Shell 的 DML 操作向《王者荣耀》游戏玩家信息表中添加数据、查询数据、删除数据。

1. *添加数据*

在 HBase Shell 中，使用 put 命令可以向表中添加数据。

【命令】

```
put <table>,<rowkey>,<family:qualifier>,<value>,<timestamp>
```

其中，timestamp 可以省略，HBase 会提供默认值，一般是时间服务器的当前时间。

示例 9

向《王者荣耀》游戏玩家信息表 gamer 中添加数据。

输入命令：

```
hbase(main):013:0> put 'gamer','row-0001','personalInfo:nickname','QGhappy.Snow'
```

输出结果：

```
0 row(s) in 0.8290 seconds
```

本示例演示的是给 gamer 表 Row Key 为“row-0001”的 personalInfo 列族中添加了列 nickname（昵称），单元值为“QGhappy.Snow”。同样也可以向 recordInfo 和 assetsInfo 列族中添加数据，代码如下。

在 recordInfo 列族中添加 ranking（排名）列，并设置单元值。

```
hbase(main):014:0> put 'gamer','row-0001','recordInfo:ranking','one'
0 row(s) in 0.8290 seconds
```

在 assetsInfo 列族中添加 integral（积分）列，并设置单元值。

```
hbase(main):015:0> put 'gamer','row-0001','assetsInfo: integral','10000'
0 row(s) in 0.8290 seconds
```

在同一个列族中可以添加多个列。如下代码所示是向 personalInfo 列族添加 gameID 列。

```
hbase(main):016:0> put 'gamer','row-0001','personalInfo:gemeID','00000000'
0 row(s) in 0.5500 seconds
```

前面的操作相当于插入了一行数据，行键为“row-0001”，还可以重新指定行键，进行数据的插入，代码如下。

```
hbase(main):017:0> put 'gamer','row-0002','personalInfo:nickname','XQMaster'
0 row(s) in 0.8290 seconds
hbase(main):018:0> put 'gamer','row-0002','recordInfo:ranking','two'
0 row(s) in 0.8290 seconds
hbase(main):019:0> put 'gamer','row-0002','assetsInfo: integral','10000'
0 row(s) in 0.8290 seconds
hbase(main):020:0> put 'gamer','row-0002','personalInfo:gemeID','11111111'
0 row(s) in 0.5500 seconds
```

2. **查询数据**

前面已经介绍了如何向表中插入数据，接下来介绍如何从表中查询数据。

（1）scan（扫描表）

【命令】

```
scan <table>, {COLUMNS => [ <family:qualifier>,.... ], LIMIT => num}
```

其中，大括号中的内容为扫描条件，如果不指定就查询所有数据。

示例 10

扫描 gamer 表中的数据。

输入命令：

```
hbase(main):021:0> scan 'gamer'
```

输出结果：

```
ROW                  COLUMN+CELL
row-0001             column=assetsInfo: integral, timestamp=1531371272178, value=10000
row-0001             column=personalInfo:gemeID, timestamp=1531371278217, value=00000000
row-0001             column=personalInfo:nickname, timestamp=1531371261621, value=QGhappy.
                     Snow
```

```
row-0001        column=recordInfo:ranking, timestamp=1531371267141, value=one
row-0002        column=assetsInfo: integral, timestamp=1531371617663, value=10000
row-0002        column=personalInfo:gemeID, timestamp=1531371622930, value=11111111
row-0002        column=personalInfo:nickname, timestamp=1531371594539, value=XQMaster
row-0002        column=recordInfo:ranking, timestamp=1531371607531, value=two
2 row(s) in 0.0310 seconds
```

结果中显示共 2 行数据，因为在扫描结果中，会将相同行键的所有单元视为一行。有时，一些列族不需要显示，便可使用 scan 命令来指定需要显示的列族。

示例 11

扫描 gamer 表中 personalInfo 列族的数据。

输入命令：

```
hbase(main):022:0> scan 'gamer',{COLUMNS => 'personalInfo'}
```

输出结果：

```
ROW             COLUMN+CELL
row-0001        column=personalInfo:gemeID, timestamp=1531371278217, value=00000000
row-0001        column=personalInfo:nickname, timestamp=1531371261621, value=QGhappy.
                Snow
row-0002        column=personalInfo:gemeID, timestamp=1531371622930, value=11111111
row-0002        column=personalInfo:nickname, timestamp=1531371594539, value=XQMaster
2 row(s) in 0.0260 seconds
```

使用同样的方式可以查看 gamer 表中 recordInfo 和 assetsInfo 列族的数据。除了可以指定列族外，还可以指定列键来扫描。扫描所有行的列键为“personalInfo:nickname”的单元的代码的如下所示。

输入命令：

```
hbase(main):023:0> scan 'gamer',{COLUMN => ['personalInfo:nickname']}
```

输出结果：

```
ROW          COLUMN+CELL
row-0001     column=personalInfo:nickname,timestamp=1531371261621,value=QGhappy.Snow
row-0002     column=personalInfo:nickname, timestamp=1531371594539, value=XQMaster
2 row(s) in 0.0940 seconds
```

从输入命令可以看出，指定列键和列族命令的不同之处是将 COLUMNS 替换为 COLUMN，表示当前扫描的是列键，注意区分大小写。

（2）get（获取数据）

get 命令用于获取行的所有单元或者某个指定的单元。

【命令】

get '表名称', '行键',{COLUMNS=>['列族名 1', '列族名 2', …],参数名=>参数值…}

get '表名称', '行键',{COLUMN=>['列键 1', '列键 2', …],参数名=>参数值…}

与 scan 命令相比，get 命令多了一个行键参数。scan 查找的目标是全表的某个列族、列键，而 get 查找的目标是某行的某个列族、列键。

示例 12

在 gamer 表中查找行键为“row-0001”的所有单元。

输入命令：

```
hbase(main):024:0> get 'gamer','row-0001'
```

输出结果：

```
COLUMN                      CELL
assetsInfo: integral        timestamp=1531371272178, value=10000
personalInfo:gemeID         timestamp=1531371278217, value=00000000
personalInfo:nickname       timestamp=1531371261621, value=QGhappy.Snow
recordInfo:ranking          timestamp=1531371267141, value=one
4 row(s) in 0.0340 seconds
```

从输出结果可以看出，不指定列族或列键，会输出对应行键的所有列键单元。

示例 13

查找 gamer 表中行键为“row-0001”，列键为“personalInfo:nickname”的单元。

输入命令：

```
hbase(main):025:0> get 'gamer','row-0001', {COLUMN=>'personalInfo:nickname'}
```

输出结果：

```
COLUMN                      CELL
personalInfo:nickname       timestamp=1531371261621, value=QGhappy.Snow
1 row(s) in 0.0350 seconds
```

上面的命令也可以简写为 get 'gamer','row-0001', 'personalInfo:nickname'，读者可以自行验证。

示例 14

查找 gamer 表中行键为“row-0001”，列族为“personalInfo”的所有单元。

输入命令：

```
hbase(main):026:0> get 'gamer','row-0001', {COLUMNS=>'personalInfo'}
```

输出结果：

```
COLUMN                      CELL
personalInfo:gemeID         timestamp=1531371278217, value=00000000
personalInfo:nickname       timestamp=1531371261621, value=QGhappy.Snow
2 row(s) in 0.0330 seconds
```

3．*删除数据*

（1）delete

使用 delete 命令可以删除 HBase 中的一个单元。

【命令】

delete '表名称', '行键', '列键'

示例 15

删除 gamer 表中行键为“row-0001”，列键为“personalInfo:nickname”的所有单元。

输入命令：

```
hbase(main):027:0> delete 'gamer','row-0001','personalInfo:nickname'
```

输出结果：

```
0 row(s) in 0.1460 seconds
```

通过 get 命令查看删除是否成功。

```
hbase(main):028:0> get 'gamer','row-0001','personalInfo'
COLUMN                          CELL
personalInfo:gemeID             timestamp=1531371278217, value=00000000
1 row(s) in 0.0320 seconds
```

从结果可以看出，行键为“row-0001”，列键为“personalInfo:nickname”的单元已经不存在，表示删除成功。

（2）deleteall

使用 deleteall 命令可以删除一行。

【命令】

```
deleteall '表名称', '行键'
```

示例 16

删除 gamer 表中行键为“row-0001”的一行数据。

输入命令：

```
hbase(main):029:0> deleteall 'gamer','row-0001'
```

输出结果：

```
0 row(s) in 0.1740 seconds
```

通过 get 命令查看删除是否成功。

```
hbase(main):030:0> get 'gamer','row-0001'
COLUMN                          CELL
0 row(s) in 0.0510 seconds
```

从结果可以看出，行键为“row-0001”的单元数据已经不存在，表示删除成功。

（3）truncate

使用 truncate 命令可以删除表中的所有数据。

【命令】

```
truncate '表名称'
```

示例 17

清空 gamer 表中的所有数据。

输入命令：

```
hbase(main):031:0> truncate 'gamer'
```

输出结果：

```
Truncating 'gamer' table (it may take a while):
- Disabling table...
- Truncating table...
0 row(s) in 7.0800 seconds
```

使用 scan 命令验证删除是否成功。

```
hbase(main):032:0> scan 'gamer'
ROW                  COLUMN+CELL
```

```
0 row(s) in 0.0510 seconds
```

从结果中可以看出，gamer 表中已经没有数据，证明 truncate 命令执行成功。

更多有关 HBase Shell 的访问操作请扫描二维码。

HBase Shell 操作

7.1.3　技能实训

使用 HBase Shell 完成《英雄联盟》游戏玩家信息表的管理操作。

关键步骤：

（1）使用 create 命令创建《英雄联盟》游戏玩家信息表 lol，列族分别为 personalInfo（个人信息）、recordInfo（战绩信息）以及 assetsInfo（资产信息）。

（2）使用 put 命令向 lol 表中添加数据，列键自己指定。

（3）使用 get 命令查看 lol 表中的数据（指定列键、指定列族）。

（4）使用 delete 命令删除数据。

任务 2　使用 HBase Java API 完成《王者荣耀》游戏玩家信息管理操作

【任务描述】

本任务使用 HBase 提供的 Java API 实现对《王者荣耀》游戏玩家信息表的操作。

【关键步骤】

（1）搭建 HBase 开发环境。

（2）使用 Java API 完成《荣耀游戏》玩家信息表的管理操作。

7.2.1　开发环境搭建

HBase 除了可以使用 HBase Shell 的方式来操作外，还可以通过 Java API 的方式来访问。HBase 的 Java API 允许在 IDE 环境下对 HBase 进行编程。本节使用 Java API 的方式来完成对《王者荣耀》游戏玩家信息表 gamer 的操作，使用的开发工具是 IntelliJ IDEA + Maven。

1．使用 IntelliJ IDEA+Maven 搭建开发环境

（1）在 IDEA 中创建 maven 项目 hbaseJavaAPI。

（2）添加 Maven 依赖包。Maven pom 文件的 jar 包依赖如下。

```
<dependencies>
    <dependency>
        <groupId>org.apache.hbase</groupId>
        <artifactId>hbase-client</artifactId>
        <version>1.2.0-cdh5.14.2</version>
    </dependency>
```

```
<dependency>
    <groupId>org.apache.hbase</groupId>
    <artifactId>hbase-common</artifactId>
    <version>1.2.0-cdh5.14.2</version>
</dependency>
<dependency>
    <groupId>org.apache.hbase</groupId>
    <artifactId>hbase-protocol</artifactId>
    <version>1.2.0-cdh5.14.2</version>
</dependency>
<dependency>
    <groupId>org.apache.hadoop</groupId>
    <artifactId>hadoop-common</artifactId>
    <version>2.6.0-cdh5.14.2</version>
</dependency>
<dependency>
    <groupId>org.apache.hadoop</groupId>
    <artifactId>hadoop-hdfs</artifactId>
    <version>2.6.0-cdh5.14.2</version>
</dependency>
<dependency>
    <groupId>commons-logging</groupId>
    <artifactId>commons-logging</artifactId>
    <version>1.2</version>
</dependency>
<dependency>
    <groupId>log4j</groupId>
    <artifactId>log4j</artifactId>
    <version>1.2.17</version>
</dependency>
<dependency>
    <groupId>junit</groupId>
    <artifactId>junit</artifactId>
    <version>4.11</version>
</dependency>
<dependency>
    <groupId>com.google.guava</groupId>
    <artifactId>guava</artifactId>
    <version>15.0</version>
</dependency>
<dependency>
    <groupId>com.yammer.metrics</groupId>
    <artifactId>metrics-core</artifactId>
    <version>2.2.0</version>
</dependency>
```

```
<dependency>
    <groupId>commons-collections</groupId>
    <artifactId>commons-collections</artifactId>
    <version>3.2.2</version>
</dependency>
<dependency>
    <groupId>org.slf4j</groupId>
    <artifactId>slf4j-api</artifactId>
    <version>1.7.25</version>
</dependency>
<dependency>
    <groupId>org.slf4j</groupId>
    <artifactId>slf4j-log4j12</artifactId>
    <version>1.7.25</version>
</dependency>
<dependency>
    <groupId>log4j</groupId>
    <artifactId>log4j</artifactId>
    <version>1.2.17</version>
</dependency>
<dependency>
    <groupId>commons-configuration</groupId>
    <artifactId>commons-configuration</artifactId>
    <version>1.10</version>
</dependency>
<dependency>
    <groupId>org.apache.hadoop</groupId>
    <artifactId>hadoop-auth</artifactId>
    <version>3.1.0</version>
</dependency>
<dependency>
    <groupId>commons-cli</groupId>
    <artifactId>commons-cli</artifactId>
    <version>1.2</version>
</dependency>
<dependency>
    <groupId>com.google.protobuf</groupId>
    <artifactId>protobuf-java</artifactId>
    <version>2.5.0</version>
</dependency>
<dependency>
    <groupId>org.apache.avro</groupId>
    <artifactId>avro</artifactId>
    <version>1.7.6-cdh5.14.2</version>
</dependency>
```

```
    <dependency>
        <groupId>org.cloudera.htrace</groupId>
        <artifactId>htrace-core</artifactId>
        <version>2.04</version>
    </dependency>
    <dependency>
        <groupId>org.apache.htrace</groupId>
        <artifactId>htrace-core</artifactId>
        <version>3.1.0-incubating</version>
    </dependency>
    <dependency>
        <groupId>org.apache.htrace</groupId>
        <artifactId>htrace-core4</artifactId>
        <version>4.0.1-incubating</version>
    </dependency>
    <dependency>
        <groupId>io.netty</groupId>
        <artifactId>netty-all</artifactId>
        <version>4.1.25.Final</version>
    </dependency>
</dependencies>
```

2. 编写单元测试方法

本节采用单元测试的方式来演示 Java API 操作 HBase。将 HBase 的连接以及表的实例对象的初始化操作放在单元测试初始化方法中，关闭操作放在单元测试关闭方法中。

示例 18

编写操作 gamer 表的初始化方法。

关键代码：

```
import java.io.IOException;
import org.apache.hadoop.conf.Configuration;
import org.apache.hadoop.hbase.HBaseConfiguration;
import org.apache.hadoop.hbase.client.*;
import org.apache.hadoop.hbase.client.Connection;
import org.apache.hadoop.hbase.client.ConnectionFactory;
import org.apache.hadoop.hbase.*;
import org.junit.After;
import org.junit.Before;
import org.junit.Test;
import org.apache.hadoop.hbase.filter.CompareFilter.CompareOp;
import org.apache.hadoop.hbase.filter.SingleColumnValueFilter;
import org.apache.hadoop.hbase.filter.Filter;
import org.apache.hadoop.hbase.util.Bytes;
public class HbaseAPIOperations {
    static Configuration configuration = null;
```

```
private Connection connection = null;
private Table table = null;
 @Before
 public void init(){
     configuration = HBaseConfiguration.create();
     configuration.set("hbase.zookeeper.quorum", "192.168.85.239:2181");
     try {
         connection = ConnectionFactory.createConnection(configuration);
         table = connection.getTable(TableName.valueOf("gamer"));
     } catch (IOException e) {
         e.printStackTrace();
     }
 }
@After
public void close() throws Exception{
    table.close();
    connection.close();
}
}
```

7.2.2 核心 API

HBase 开发环境准备好以后，就可以使用 HBase 核心 API 来完成对 gamer 表的管理操作。

1. HBase Java 核心 API 介绍

HBase Java API 提供了对应于 HBase 数据模型的核心类，其对应关系如表 7-4 所示。

表 7-4　HBase Java API 与 HBase 数据模型对应表

HBase Java 类	HBase 数据模型
HBaseConfiguration	HBase 数据库
Connection	
Admin	
HTableDescriptor	表（Table）
TableName	
Table	
HColumnDescriptor	列族（Column Family）
Put	添加数据（put 命令）
Get	获取数据（get 命令）
Result	存储 Get 或 Scan 操作后获取的单行的值
ResultScanner、Scan	扫描表（scan 命令）

从表 7-4 中可以看出，HBase 提供了很多 API，下面分别介绍。

（1）HBaseConfiguration

org.apache.hadoop.hbase.HBaseConfiguration 类是 HBase 的配置类，通过 HBase

Configuration 类可以对 HBase 运行时环境进行配置。所有的操作都需要先创建 HBaseConfiguration 类的实例，可以通过该类的静态方法 create()获得。用法示例如下。

```
Configuration configuration = HBaseConfiguration.create();
```

上述语句会从 classpath 中查找 hbase-site.xml 文件，如果找不到，则使用默认配置文件 hbase-default.xml。

（2）Connection

org.apache.hadoop.hbase.client.Connection 接口表示 HBase 的连接，通过 Connection 接口的实例能够查找到 HMaster、定位 HRegion 在集群中的位置并缓存，另外，Table 和 Admin 实例也要从 Connection 实例中获取数据。可以参考示例 18 初始化方法中的代码。

（3）Admin

org.apache.hadoop.hbase.client.Admin 是为管理 HBase 表提供的接口，它提供了相当于 HBase Shell 中的 DDL 操作。具体的方法如表 7-5 所示。

表 7-5 Admin 接口常用方法

返回值	方法	功能
void	createTable(HTableDescriptor desc)	创建一个新表
HTableDescriptor[]	listTables()	列出所有的表
void	deleteTable(TableName tableName)	删除一个已经存在的表
void	enableTable(TableName tableName)	使表有效
void	disableTable(TableName tableName)	使表无效
void	modifyTable(TableName tableName,HTableDescriptor htd)	修改一个已经存在的表
boolean	TableExists(TableName tableName)	检查表是否存在

（4）HTableDescriptor、TableName、Table

org.apache.hadoop.hbase.HTableDescriptor 类包含了表的名字及对应的列族。

org.apache.hadoop.hbase.TableName 类是对应表名的封装类。表名的完整形式为：<table namespace>:<table qualifier>，如果没有指定表名字空间，会使用默认的 default，意味着使用 TableName 类可以获得表的名字空间和表名。而定义表名可以使用 TableName.valueof(String tableName)方法。

org.apache.hadoop.hbase.client.Table 接口主要用于和 HBase 中的表进行通信，使用 Connection 的 getTable(TableName tablename)方法可以获取 Table 接口的实例，该接口可以获取、添加、删除、扫描 HBase 表中的数据。Table 接口的常用方法如表 7-6 所示。

表 7-6 Table 接口常用方法

返回值	方法	功能
void	put(Put put)	向表添加值，put 表示添加操作
Result	get(Get get)	获取单元值，get 表示获取操作
void	delete(Delete delete)	删除指定的单元/行，delete 表示删除操作

续表

返回值	方法	功能
ResultScanner	getScanner(Scan)	获取当前表的给定列族的 Scanner 实例，ResultScanner 代表结果列表
	getScanner(byte[] family,byte[] qualifier)	
	getScanner(byte family)	
boolean	exists(Get get)	检查 Get 实例所对应的值是否在 Table 中
HTableDescriptor	getTableDescriptor()	获取表的 HTableDescriptor 实例
TableName	getName()	获取表名

（5）HColumnDescriptor

org.apache.hadoop.hbase.HColumnDescriptor 类是对列族的描述，包含了列族名称、版本号、压缩设置等信息。该类的实例只在创建表或给表添加、删除列族时使用。需注意的是，一旦列族被删除，对应列族中所保存的数据也将同时被删除。

（6）Put

org.apache.hadoop.hbase.client.Put 类为指定的行键添加列键和值。在 HBase Shell 操作中，使用 put 命令添加数据时，有四个参数：表名、行键、列键和值。在 Java 中，表名由 Table 实例传递，其余三个参数由 Put 实例提供，其中，行键由构造方法指定，列键和值由对应的方法设置。Put 类的构造方法如下。

➢ Put(byte[] row)：对应指定行。row 为行键的字节码，通常使用 org.apache.hadoop.hbase.util.Bytes 类的 toBytes(String rowKey)方法获取。

➢ Put(byte[] row,long ts)：创建 Put 实例并手动指定时间戳。

Put 类的常用方法如表 7-7 所示。

表 7-7 Put 类常用方法

返回值	方法	功能
Put	addColumn(byte[]family, byte[] qualifier,byte[] value)	添加列键和值
Put	addColumn(byte[] family, byte[] qualifier,long ts,byte[] value)	添加列键和值并指定时间戳
List<Cell>	get(byte[] family,byte[] qualifier)	返回 Put 实例中与指定列键匹配的项，Cell 是单元格实例
boolean	has(byte[] family,byte[] qualifier)	检查 Put 实例中是否包含指定的列键

（7）Get

Get 类与 HBase Shell 中的 get 命令功能类似，用来获取表中单个行的数据。Get 类的常用方法如表 7-8 所示。

表 7-8 Get 类常用方法

返回值	方法	功能
Get	addColumn(byte[] family,byte)[] qualifier)	指定列族和对应的列
Get	addFamily(byte[] family)	指定列族
Get	setTimeRange(long minStamp,long maxStamp)	指定列的版本号区间

（8）Result

org.apache.hadoop.hbase.client.Result 类代表一行数据，可以从 Result 实例中获取单元格的行键、列键、版本号。Result 类的常用方法如表 7-9 所示。

表 7-9　Result 类常用方法

返回值	方法	功能
NavigableMap<byte[],byte[]>	getFamilyMap(byte[] family)	获取列族下所有列名与值的映射
byte[]	value()	返回第一个列键的值
byte[]	getValue(byte[] family,byte[] qualifier)	获取指定列键的值
boolean	contains(byte[] family,byte[] qualifier)	检查列键是否存在
byte[]	getRow()	返回行键
List<Cell>	listCells()	获取指定行的所有 Cell

（9）Scan、ResultScanner

在 HBase Shell 中，scan 命令可以按全表扫描，也可以按列族和列键扫描。在 Java 中，这个功能由 org.apache.hadoop.hbase.client.Scan 类实现。Scan 类默认进行全表扫描，通过其方法也可以实现按列族或按列键扫描。Scan 类的常用方法如表 7-10 所示。

表 7-10　Scan 类的常用方法

返回值	方法	功能
Scan	addColumn(byte[] family,byte[] qualifier)	获取指定列族和列修饰符对应的列
Scan	addColumn(byte[] family)	获取指定列族下的所有列

Scan 操作的扫描结果用 org.apache.hadoop.hbase.client.ResultScanner 接收，可以将其看成是多个 Result 的集合，也就是说 ResultScanner 中存放了多行数据。

2. 使用 HBase Java 核心 API 管理 gamer 表

前面介绍了 HBase Java API 提供的核心类及其功能。接下来使用实际案例来演示如何使用前面介绍的 Java 类完成对 HBase 数据库的管理以及对 gamer 表的管理。

示例 19

创建单元测试的 init 和 close 方法。

关键代码：

```
import java.io.IOException;
import org.apache.hadoop.conf.Configuration;
import org.apache.hadoop.hbase.HBaseConfiguration;
import org.apache.hadoop.hbase.client.*;
import org.apache.hadoop.hbase.client.Connection;
import org.apache.hadoop.hbase.client.ConnectionFactory;
import org.apache.hadoop.hbase.*;
import org.junit.After;
import org.junit.Before;
import org.junit.Test;
import org.apache.hadoop.hbase.filter.CompareFilter.CompareOp;
```

```
import org.apache.hadoop.hbase.filter.SingleColumnValueFilter;
import org.apache.hadoop.hbase.filter.Filter;
import org.apache.hadoop.hbase.util.Bytes;
public class HbaseAPIOperations {
    static Configuration configuration = null;
    private Connection connection = null;
    private Table table = null;
     @Before
     public void init(){
         configuration = HBaseConfiguration.create();
         configuration.set("hbase.zookeeper.quorum", "192.168.85.239:2181");
         try {
             connection = ConnectionFactory.createConnection(configuration);
             table = connection.getTable(TableName.valueOf("gamer"));
         } catch (IOException e) {
             e.printStackTrace();
         }
     }
    public static String tableName="gamer";
    /**
     * 创建表
     */
    @Test
    public void createTable() throws IOException {
        String familyName = "personalInfo,recordInfo,assetsInfo";
        HBaseAdmin admin = (HBaseAdmin) connection.getAdmin();
        if (admin.tableExists(tableName)) {
            System.out.println(tableName + " exists!");
        } else {
            String[] columnFamilyArray = familyName.split(",");
            HColumnDescriptor[] hColumnDescriptor =
                    new HColumnDescriptor[columnFamilyArray.length];
            for (int i = 0; i < hColumnDescriptor.length; i++) {
                hColumnDescriptor[i] = new HColumnDescriptor(columnFamilyArray[i]);
            }
            HTableDescriptor familyDesc = new HTableDescriptor(TableName.valueOf(tableName));
            for (HColumnDescriptor columnDescriptor : hColumnDescriptor) {
                familyDesc.addFamily(columnDescriptor);
            }
            HTableDescriptor tableDesc =
                        new HTableDescriptor(TableName.valueOf(tableName), familyDesc);
            admin.createTable(tableDesc);
            System.out.println(tableName + " create successfully!");
        }
    }
    /**
     * 添加行列数据
```

```
 */
  @Test
  public void insertData() throws IOException {
     System.out.println("start insert data ......");
     String rowKey="row-0001";
     String familyName="personalInfo";
     String qualifier="nickname";
     String value="lisi";
     // 一个 Put 代表一行数据，再新建一个 Put 表示第二行数据，每行一个唯一的 Row Key，
     // 此处的 rowKey 为 Put 构造方法中传入的值
     Put put = new Put(rowKey.getBytes());
     //本行数据的第一列
     put.addColumn(familyName.getBytes(), qualifier.getBytes(), value.getBytes());
     try {
            table.put(put);
     } catch (IOException e) {
            e.printStackTrace();
     }
}
/**
 * 删除行
 */
@Test
public void deleteRow() {
      String rowkey="row-0001";
      try {
            Delete d1 = new Delete(rowkey.getBytes());
            table.delete(d1);
            System.out.println("删除行成功!");
      } catch (IOException e) {
            e.printStackTrace();
      }
}
/**
 * 查询所有数据
 */
@Test
public void queryAll() throws Exception {
      try {
            ResultScanner rs = table.getScanner(new Scan());
            for (Result r : rs) {
                  System.out.println("获得到 rowkey:" + new String(r.getRow()));
                  for (Cell keyValue : r.rawCells()) {
                        System.out.println("列： " + new String(CellUtil.cloneFamily(keyValue))+
                                             ":"+new String(CellUtil.cloneQualifier(keyValue)) + "====
                                             值:" + new String(CellUtil.cloneValue (keyValue)));
                  }
```

```
            }
            rs.close();
        } catch (IOException e) {
            e.printStackTrace();
        }
}
/**
 * 根据 Row Key 查询
 */
@Test
public void queryByRowId() throws Exception {
        String rowId="row-0001";
        try {
            Get get = new Get(rowId.getBytes());// 根据 rowkey 查询
            Result r = table.get(get);
            System.out.println("获得到 rowkey:" + new String(r.getRow()));
            for (Cell keyValue : r.rawCells()) {
                System.out.println("列： " + new String(CellUtil.cloneFamily(keyValue))+":"+new
                                String(CellUtil.cloneQualifier(keyValue)) + "====值:" + new
                                String(CellUtil.cloneValue(keyValue)));
            }
        } catch (IOException e) {
            e.printStackTrace();
        }
}
/**
 * 根据列条件查询
 */
@Test
public void queryByCondition() {
        String familyName = "personalInf";
        String qualifier="name";
        String value="lisi";
        try {
            //当列 familyName 的值为 value 时进行查询
            Filter filter =new SingleColumnValueFilter(Bytes.toBytes(familyName), Bytes.toBytes
                                (qualifier), CompareOp. EQUAL,Bytes.toBytes
                                (value));
            Scan s = new Scan();
            s.setFilter(filter);
            ResultScanner rs = table.getScanner(s);
            for (Result r : rs) {
                System.out.println("获得到 rowkey:" + new String(r.getRow()));
                for (Cell keyValue : r.rawCells()) {
                    System.out.println("列： " + new String(CellUtil.cloneFamily(keyValue))+
                                ":"+new String(CellUtil.cloneQualifier(keyValue)) + "====
                                值:" + new String(CellUtil.clone Value(keyValue)));
```

```
                }
            }
            rs.close();
        } catch (Exception e) {
            e.printStackTrace();
        }
    }
    @After
    public void close() throws Exception{
        table.close();
        connection.close();
    }
}
```

7.2.3 技能实训

使用 Java API 的方式完成《英雄联盟》游戏玩家信息表的管理操作。

关键步骤：

（1）创建 Maven 项目（添加 pom jar 依赖、配置单元测试并初始化 Connection）。

（2）使用 Java API 创建 lol 表。

（3）使用 Java API 向 lol 表中添加数据。

（4）使用 Java API 查看 lol 表中的数据。

（5）使用 Java API 删除 lol 表中的数据。

任务 3 使用 HBase Rest API 访问《王者荣耀》游戏玩家信息表

【任务描述】

本任务主要使用 HBase Rest API 的方式实现对《王者荣耀》游戏玩家信息表的远程访问操作。

【关键步骤】

（1）启动 HBase Rest 服务。

（2）使用 Rest 接口访问 HBase。

7.3.1 启动/停止 Rest 服务命令

HBase 提供了 Rest 接口供 Java 语言之外的方式进行访问。启动 Rest 服务后，该服务上会运行一个 HBase 客户端实例，它负责协调 Rest 请求与 HBase 之间的双向交互，由于需要代理请求和响应工作，所以，使用 Rest 接口比直接使用 Java 要慢很多。

（1）启动

启动 Rest 服务有两种方式。

① 直接启动

```
$HBASE_HOME/bin/hbase rest start -p <port>
```

② 守护进程的方式启动

```
$HBASE_HOME/bin/hbase-daemon.sh start rest -p <port>
```

这两种方式都将启动一个服务器实例，在不指定端口启动的情况下，默认端口是 8080，客户端可以使用 HTTP 的方式进行访问。

示例 20

以守护进程的方式在 HBase 中启动 rest 服务。

输入命令：

```
$HBASE_HOME/bin/hbase-daemon.sh start rest -p 8585
```

启动后，使用 jps 命令可以看到，进程列表中会增加 RESTServer 进程。也可以在浏览器中输入 http://hadoop:8585/进行访问，其中“hadoop”为 HBase Master 节点的机器名，“8585”为启动 Rest 服务时指定的端口号。由这个地址返回的结果是 HBase 中所有的表。

（2）关闭

关闭 Rest 服务可以使用下面的命令。

```
$HBASE_HOME/bin/hbase-daemon.sh stop rest
```

7.3.2 访问方式

Rest 服务启动以后，就可以通过 HTTP 的方式与 HBase 进行交互。本节采用在 Chrome 浏览器中输入 URL 的方式进行演示。

示例 21

查看搭建的集群中使用的 HBase 版本。

在浏览器中输入 http://hadoop:8585/version/cluster。

输出结果：

```
1.2.0-cdh5.14.2
```

输出的版本正是本书选择的 HBase 的版本。

示例 22

查看 HBase 中 gamer 表的 schema 信息。

在浏览器中输入 http://hadoop:8585/gamer/schema。

输出结果：

{NAME => 'assetsInfo', BLOOMFILTER => 'ROW', VERSIONS => '1', IN_MEMORY => 'false', KEEP_DELETED_CELLS => 'FALSE', DATA_BLOCK_ENCODING => 'NONE', TTL => 'FOREVER', COMPRESSION => 'NONE', MIN_VERSIONS => '0', BLOCKCACHE => 'true', BLOCKSIZE => '65536', REPLICATION_SCOPE => '0'}

{NAME => 'personalInfo', BLOOMFILTER => 'ROW', VERSIONS => '1', IN_MEMORY => 'false', KEEP_DELETED_CELLS => 'FALSE', DATA_BLOCK_ENCODING => 'NONE', TTL => 'FOREVER', COMPRESSION => 'NONE', MIN_VERSIONS => '0', BLOCKCACHE => 'true', BLOCKSIZE => '65536', REPLICATION_SCOPE => '0'}

{NAME => 'recordInfo', BLOOMFILTER => 'ROW', VERSIONS => '1', IN_MEMORY => 'false', KEEP_DELETED_CELLS => 'FALSE', DATA_BLOCK_ENCODING => 'NONE', TTL => 'FOREVER',

COMPRESSION => 'NONE', MIN_VERSIONS => '0', BLOCKCACHE => 'true', BLOCKSIZE => '65536', REPLICATION_SCOPE => '0'}

从结果中可以看出，使用该 URL 可以查看表的信息，与 HBase Shell 中的 describe 命令查看到的信息类似。

通过 Rest 接口，还可以进行查看数据、创建表、删除表等操作，但是由于执行效率的问题，以及拼写 URL 比较烦琐等原因，这种方式用得较少，这里不再赘述。更多有关 Rest 的访问操作请扫描二维码。

HBase Rest 操作

7.3.3 技能实训

使用 Rest 的方式访问《王者荣耀》游戏玩家信息表。

关键步骤：

（1）在 HBase 环境中启动 Rest 服务。

（2）使用 HTTP 方式查看 HBase 软件安装版本。

（3）使用 HTTP 方式查看 lol 表的 schema 信息。

本章总结

➢ HBase 提供了 Shell 命令的方式来对数据库及表进行基本的管理，常用的命令包括 create、list、describe、put、get、delete、disable、enable、drop 等。

➢ HBase 提供了 Java API 的方式来对数据库及表进行基本的管理，核心 API 包括 HBaseConnection、Connection、Table、Admin、HTableDescriptor、HColumnDescriptor、TableName、Put、Get、Result、ResultScanner、Scan 等。

➢ 不论是采用 HBase Shell 还是 Java API 的方式，对于表单元的修改操作实质上是增加操作。HBase 保留了单元的多个版本，默认查询返回的是最新的版本。

本章作业

编码题

1．创建学生成绩表 scores，列族为 grade（年级）、course（课程），学生姓名 name 作为行键。

2．查看创建的表的详细信息。

3．向 scores 表中增加一些数据，其中 course 列族要求有 math（数学）、art（艺术）两列。数据格式如表 7-11 所示。

表 7-11 学生成绩表 scores

name	grade	course	
		math	art
John	1	84	87
Jack	2	100	89

4．全表扫描 scores 中的数据。

5．获取 Jack 的数学成绩。

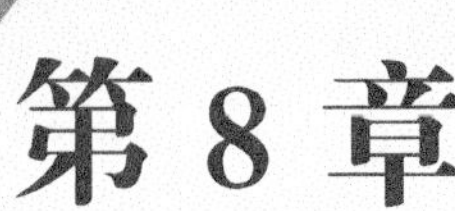

第8章

HBase应用

技能目标

- 学会管理 HBase 表空间
- 学会对 HBase 表进行权限管理
- 理解 HRegion 切分
- 理解 HBase 中的 Compaction 过程

本章任务

任务1　使用表空间管理《王者荣耀》游戏玩家信息表
任务2　对《王者荣耀》游戏玩家信息表进行权限管理
任务3　理解 HRegion 切分
任务4　了解 HBase 中的 Compaction 过程

本章资源下载

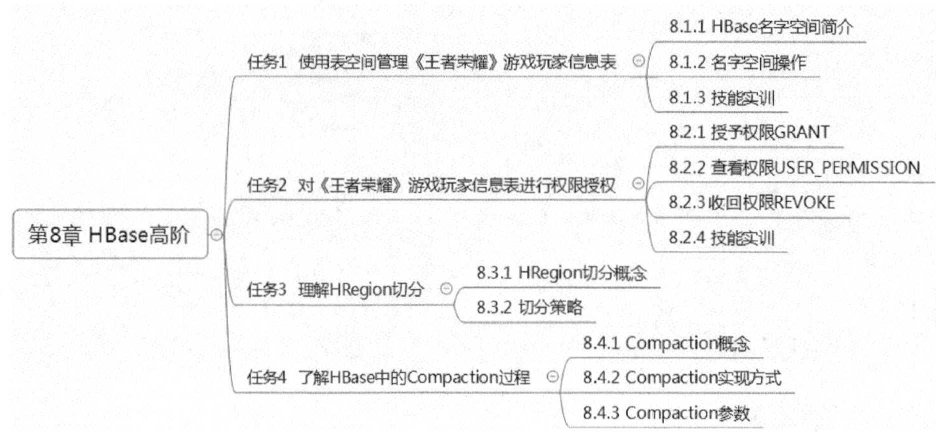

在 HBase 中，可以使用名字空间（namespace）对 HBase 中的表进行权限管理，还可以通过 HRegion 切分和 Compaction 机制对 HBase 进行优化。本章主要介绍 HBase 的名字空间、权限管理、HRegion 切分以及 Compaction 的概念。

任务 1 使用表空间管理《王者荣耀》游戏玩家信息表

【任务描述】

学会使用表空间管理《王者荣耀》游戏玩家信息表。

【关键步骤】

（1）了解 HBase 的名字空间。

（2）使用命令管理 HBase 名字空间。

8.1.1 HBase 名字空间简介

在 HBase 中，名字空间（namespace），也称表空间、命名空间。是对一组表的逻辑分组，方便对表进行业务上的划分，类似于关系型数据库中数据库（database）的概念。HBase 从 0.98.0 和 0.95.2 两个版本开始支持名字空间级别的授权操作，HBase 全局管理员可以创建、修改和回收名字空间的授权。在没有名字空间的概念之前，对 HBase 表的管理主要通过人为对 HBase 表的表名进行规则设置来实现。

HBase 中有两个系统内置的预定义的名字空间。

（1）hbase。系统名字空间，用于包含 HBase 的内部表。

（2）default。所有未指定名字空间的表都自动进入该名字空间。

8.1.2 名字空间操作

在关系型数据库中，管理员可以对数据库进行创建、删除、修改等操作，在 HBase

中，管理员也可以对名字空间进行创建、修改、删除等操作。本节主要介绍在 HBase 中如何对 namespace 进行操作。

1. HBase Shell 操作名字空间

首先进入 HBase Shell，命令如下。

```
[hadoop@hadoop ~]$ hbase shell
```

进入到 Shell 环境后，就可以使用命令对名字空间进行操作。

（1）查看 HBase 中所有的名字空间

【命令】

```
list_namespace
```

示例 1

列出 HBase 中所有的名字空间。

关键代码：

```
hbase(main):002:0> list_namespace
NAMESPACE
default
hbase
2 row(s) in    0.0310 seconds
```

（2）创建名字空间

【命令】

```
create_namespace 'namespace 名称'
```

示例 2

创建管理《王者荣耀》游戏玩家信息表的名字空间。

关键代码：

```
hbase(main):001:0> create_namespace 'gamer_namespace'
0 row(s) in 1.9420 seconds
```

通过 list_namespace 命令可以查看是否创建成功。

关键代码：

```
hbase(main):002:0> list_namespace
NAMESPACE
default
gamer_namespace
hbase
3 row(s) in 0.9450 seconds
```

从结果中可以看出，gamer_namespace 已经创建成功。

（3）查看名字空间详细信息

【命令】

```
describe_namespace 'namespace 名称'
```

示例 3

查看示例 2 创建的 gamer_namespace 的详细信息。

```
hbase(main):003:0> describe_namespace 'gamer_namespace'
```

```
DESCRIPTION
{NAME => 'gamer_namespace'}
1 row(s) in 0.1070 seconds
```

（4）在名字空间下创建表

【命令】

```
create 'namespace:tablename', 'family',…
```

其中，namespace 为名字空间名称，tablename 为表名称，family 为列族名称。

注意

在名字空间中创建表时，要保证名字空间已经在 HBase 中存在，否则会报错。

示例 4

在示例 2 创建的 gamer_namespace 中创建《王者荣耀》游戏玩家信息表 gamer。

```
hbase(main):004:0> create 'gamer_namespace:gamer', 'personalInfo'
0 row(s) in 5.1890 seconds

=> Hbase::Table - gamer_namespace:gamer
```

（5）查看名字空间中的表

【命令】

```
list_namespace_tables 'namespace 名称'
```

示例 5

查看示例 2 中创建的 gamer_namespace 下的表。

```
hbase(main):005:0> list_namespace_tables 'gamer_namespace'
TABLE
gamer
1 row(s) in 0.3060 seconds
```

从结果中可以看出，示例 4 在 gamer_namespace 中创建的 gamer 表已经存在。

（6）删除名字空间

【命令】

```
drop_namespace 'namespace 名称'
```

示例 6

删除示例 2 创建的 gamer_namespace 。

关键代码：

```
hbase(main):006:0> drop_namespace 'gamer_namespace'
ERROR: org.apache.hadoop.hbase.constraint.ConstraintException: Only empty namespaces can be removed. Namespace gamer_namespace has 1 tables
    at org.apache.hadoop.hbase.master.TableNamespaceManager.remove(TableNamespaceManager.java:200)
    at org.apache.hadoop.hbase.master.HMaster.deleteNamespace(HMaster.java:2576)
    at org.apache.hadoop.hbase.master.MasterRpcServices.deleteNamespace(MasterRpcServices.java:496)
```

```
atorg.apache.hadoop.hbase.protobuf.generated.MasterProtos$MasterService$2.callBlockingMethod
(MasterProtos.java:55730)
    at org.apache.hadoop.hbase.ipc.RpcServer.call(RpcServer.java:2191)
    at org.apache.hadoop.hbase.ipc.CallRunner.run(CallRunner.java:112)
    at org.apache.hadoop.hbase.ipc.RpcExecutor$Handler.run(RpcExecutor.java:183)
    at org.apache.hadoop.hbase.ipc.RpcExecutor$Handler.run(RpcExecutor.java:163)

Drop the named namespace. The namespace must be empty.
```

从结果中可以看出，要删除 gamer_namespace，必须保证 gamer_namespace 是空的，也就是说，必须保证该 namespace 下不包含任何表。解决方式如下。

① 禁用 gamer 表

```
hbase(main):007:0> disable 'gamer_namespace:gamer'
0 row(s) in 2.7700 seconds
```

② 删除 gamer 表

```
hbase(main):008:0> drop 'gamer_namespace:gamer'
0 row(s) in 1.8840 seconds
```

③ 删除 gamer_namespace

```
hbase(main):009:0> drop_namespace 'gamer_namespace'
0 row(s) in 5.0820 seconds
```

表删除后，再删除名字空间就不会出现错误，可以使用 list_namespace 查看 gamer_namespace 是否删除成功。

关键代码：

```
hbase(main):010:0> list_namespace
NAMESPACE
default
hbase
2 row(s) in    0.0270 seconds
```

从列出的名字空间可以看到，gamer_namespace 已经不存在，证明已经成功删除。

2. Java API 操作名字空间

除了可以使用 HBase Shell 操作 HBase 名字空间之外，还可以使用 Java API 的方式来管理名字空间。接下来介绍使用 Java API 来管理名字空间的操作，同样是使用 IntelliJ IDEA + Maven 来开发。

（1）在 IDEA 中创建 maven 项目 namespaceJavaAPI。

（2）添加 Maven 依赖包，Maven pom 文件如下。

```
<dependencies>
    <dependency>
        <groupId>org.apache.hbase</groupId>
        <artifactId>hbase-client</artifactId>
        <version>1.2.0-cdh5.14.2</version>
    </dependency>
    <dependency>
```

```
        <groupId>org.apache.hbase</groupId>
        <artifactId>hbase-common</artifactId>
        <version>1.2.0-cdh5.14.2</version>
    </dependency>
    <dependency>
        <groupId>org.apache.hbase</groupId>
        <artifactId>hbase-protocol</artifactId>
        <version>1.2.0-cdh5.14.2</version>
    </dependency>
    <dependency>
        <groupId>org.apache.hadoop</groupId>
        <artifactId>hadoop-common</artifactId>
        <version>2.6.0-cdh5.14.2</version>
    </dependency>
    <dependency>
        <groupId>org.apache.hadoop</groupId>
        <artifactId>hadoop-hdfs</artifactId>
        <version>2.6.0-cdh5.14.2</version>
    </dependency>
    <dependency>
        <groupId>commons-logging</groupId>
        <artifactId>commons-logging</artifactId>
        <version>1.2</version>
    </dependency>
    <dependency>
        <groupId>log4j</groupId>
        <artifactId>log4j</artifactId>
        <version>1.2.17</version>
    </dependency>
    <dependency>
        <groupId>junit</groupId>
        <artifactId>junit</artifactId>
        <version>4.11</version>
    </dependency>
    <dependency>
        <groupId>com.google.guava</groupId>
        <artifactId>guava</artifactId>
        <version>15.0</version>
    </dependency>
    <dependency>
        <groupId>com.yammer.metrics</groupId>
        <artifactId>metrics-core</artifactId>
        <version>2.2.0</version>
    </dependency>
    <dependency>
```

```
        <groupId>commons-collections</groupId>
        <artifactId>commons-collections</artifactId>
        <version>3.2.2</version>
    </dependency>
    <dependency>
        <groupId>org.slf4j</groupId>
        <artifactId>slf4j-api</artifactId>
        <version>1.7.25</version>
    </dependency>
    <dependency>
        <groupId>org.slf4j</groupId>
        <artifactId>slf4j-log4j12</artifactId>
        <version>1.7.25</version>
    </dependency>
    <dependency>
        <groupId>log4j</groupId>
        <artifactId>log4j</artifactId>
        <version>1.2.17</version>
    </dependency>
    <dependency>
        <groupId>commons-configuration</groupId>
        <artifactId>commons-configuration</artifactId>
        <version>1.10</version>
    </dependency>
    <dependency>
        <groupId>org.apache.hadoop</groupId>
        <artifactId>hadoop-auth</artifactId>
        <version>3.1.0</version>
    </dependency>
    <dependency>
        <groupId>commons-cli</groupId>
        <artifactId>commons-cli</artifactId>
        <version>1.2</version>
    </dependency>
    <dependency>
        <groupId>com.google.protobuf</groupId>
        <artifactId>protobuf-java</artifactId>
        <version>2.5.0</version>
    </dependency>
    <dependency>
        <groupId>org.apache.avro</groupId>
        <artifactId>avro</artifactId>
        <version>1.7.6-cdh5.14.2</version>
    </dependency>
    <dependency>
```

```
            <groupId>org.cloudera.htrace</groupId>
            <artifactId>htrace-core</artifactId>
            <version>2.04</version>
        </dependency>
        <dependency>
            <groupId>org.apache.htrace</groupId>
            <artifactId>htrace-core</artifactId>
            <version>3.1.0-incubating</version>
        </dependency>
        <dependency>
            <groupId>org.apache.htrace</groupId>
            <artifactId>htrace-core4</artifactId>
            <version>4.0.1-incubating</version>
        </dependency>
        <dependency>
            <groupId>io.netty</groupId>
            <artifactId>netty-all</artifactId>
            <version>4.1.25.Final</version>
        </dependency>
    </dependencies>
```

（3）使用 Java API 完成名字空间的创建以及在名字空间中创建表。

示例 7

使用 Java API 完成对《王者荣耀》游戏玩家信息表 namespace 的管理。

本示例同样采用单元测试的方式来实现。

关键代码：

```
import org.apache.hadoop.conf.Configuration;
import org.apache.hadoop.hbase.*;
import org.apache.hadoop.hbase.client.Admin;
import org.apache.hadoop.hbase.client.Connection;
import org.apache.hadoop.hbase.client.ConnectionFactory;
import org.apache.hadoop.hbase.util.Bytes;
import java.io.IOException;
import org.junit.After;
import org.junit.Before;
import org.junit.Test;
public class NamespaceJavaAPI {
    static Configuration configuration = null;
    static Connection connection = null;
    //管理员对象
    static Admin admin = null;
    @Before
    public void init(){
        configuration = HBaseConfiguration.create();
        configuration.set("hbase.zookeeper.quorum", "hadoop:2181");
```

```
        try {
            connection = ConnectionFactory.createConnection(configuration);
            admin = connection.getAdmin();
        } catch (IOException e) {
            e.printStackTrace();
        }
    }
    **
    * 创建表空间
    */
    @Test
    public void createNamespace() throws IOException {
        NamespaceDescriptor namespaceDescriptor =
                        NamespaceDescriptor.create("gamer_namespace").build();
        admin.createNamespace(namespaceDescriptor);
    }
    /**
    * 创建表
    */
    @Test
    public void createTable() throws IOException {
        //表名
        TableName tableName = TableName.valueOf("gamer_namespace","gamer");
        //表描述
        HTableDescriptor desc = new HTableDescriptor(tableName);
        //列族描述
        HColumnDescriptor coldef = new HColumnDescriptor(Bytes.toBytes("personalInfo"));
        //表加入列族
        desc.addFamily(coldef);
        //创建表
        admin.createTable(desc);
    }
    /**
    * 验证表是否可用
    */
    @Test
    public void isTableAvailable() throws IOException {
        TableName tableName = TableName.valueOf("gamer_namespace","gamer");
        //校验表是否可用
        boolean avail = admin.isTableAvailable(tableName);
        System.out.println("Table available: "+avail);
    }
    @After
    public void close() throws Exception{
        connection.close();
```

```
            admin.close();
        }
    }
```

在实际工作中，名字空间一般都是在建模阶段通过 HBase Shell 的方式进行创建，使用 Java API 的方式进行创建的应用场景并不多。

注意

在使用 Java API 操作 HBase 名字空间的时候，需要将 HBase 的配置文件 hbase-site.xml 复制到工程的 classpath 中，或者通过 Configuration 对象设置相关的属性，否则程序将获取不到集群相关信息，也就无法找到集群，运行时会报错。

8.1.3 技能训练

使用 HBase Shell 的方式完成《英雄联盟》游戏玩家信息表 namespace 的管理操作。

关键步骤：

（1）创建 lol_namespace。

（2）查看 lol_namespace 的详细信息。

（3）在 lol_namespace 下创建 lol_gamer 表，使用 create 命令创建《英雄联盟》玩家信息表 lol_gamer，列族分别为 personalInfo（个人信息）、recordInfo（战绩信息）以及 assetsInfo（资产信息）。

（4）删除 lol_namespace。

任务 2 对《王者荣耀》游戏玩家信息表进行权限管理

【任务描述】

本任务使用权限管理命令对《王者荣耀》游戏玩家信息表进行权限管理操作。

【关键步骤】

（1）使用 GRANT 命令进行授权。

（2）使用 REVOKE 命令收回权限。

（3）使用 USER_PERMISSON 命令查看权限。

8.2.1 授予权限 GRANT

与关系型数据库一样，HBase 中也有用户和权限的概念。在 HBase 中，提供了五个权限标识符：RWXCA，分别对应读 READ（'R'）、写 WRITE（'W'）、执行 EXEC（'X'）、创建 CREATE（'C'）和超级管理员 ADMIN（'A'）。

HBase 提供的安全管控级别包括以下六种。

（1）SuperUser：拥有所有权限的超级管理员用户，可以通过 hbase.superuser 参数配置。

（2）Global：全局权限，可以作用在集群所有的表上。

（3）namcspacc：名字空间级别权限。

（4）Table：表级别权限。

（5）ColumnFamily：列族级别权限。

（6）Cell：单元级权限。

HBase 的权限管理依赖于协处理器，需要在 HBase 配置文件 hbase-site.xml 中配置相关的属性。具体参数配置如下。

```
<property>
    <name>hbase.superuser</name>
    <value>hbase</value>
</property>
<property>
    <name>hbase.security.authorization</name>
    <value>true</value>
</property>
<property>
    <name>hbase.coprocessor.master.classes</name>
    <value>org.apache.hadoop.hbase.security.access.AccessController</value>
</property>
<property>
    <name>hbase.coprocessor.region.classes</name>
    <value>org.apache.hadoop.hbase.security.token.TokenProvider,
          org.apache.hadoop.hbase.security.access.AccessController
    </value>
</property>
```

与关系型数据库一样，在 HBase 中，权限的授予也是使用 GRANT 命令。

【命令】

```
grant <user>, <permissions>[,<@namespace> [,<table> [,<column family> [,<column qualifier>]]]]
```

使用这个命令可以给一个用户赋予权限。其中，user 为用户，permissions 为权限，其他分别对应表名、列族及列名。

示例 8

给用户 bigdata 授予对示例 4 创建的 gamer 表的列键 personalInfo:name 以 RWCA 权限。

关键代码：

```
hbase(main):006:0> grant 'bigdata','RWCA','gamer_namespace:gamer','personalInfo','name'
0 row(s) in 2.1370 seconds
```

表名使用 namespace:tablename 的方式来表示。

这里只演示列族级别的权限授予，读者可以自行完成其他级别的权限授予。

8.2.2 查看权限 USER_PERMISSION

在示例 8 中，给用户 bigdata 授予了 gamer 表的 RWCA 权限，本节使用 USER_PERMISSION 命令来查看表的权限。

【命令】

```
user_permission 'tablename'
```

其中，tablename 为表名。

示例 9

查看 gamer_namespace:gamer 表的权限。

关键代码：

```
hbase(main):008:0> user_permission 'gamer_namespace:gamer'
User                Namespace,Table,Family,Qualifier:Permission
bigdata             gamer_namespace,gamer_namespace:gamer,personalInfo,name:
[Permission:actions=READ,WRITE,CREATE,ADMIN]
```

从上面的结果可以看出，bigdata 用户对表 gamer_namespace:gamer 的 personalInfo:name 具有 READ、WRITE、CREATE、ADMIN 权限。这也验证了示例 8 的授权操作已经正确执行。

8.2.3 收回权限 REVOKE

与关系型数据库一样，在 HBase 中，收回权限也是使用 REVOKE 命令。

【命令】

```
revoke <user> [,<@namespace>[,<table> [,<column family> [,<column qualifier>]]]]
```

使用这个命令可以收回用户对表的权限。其中，user 为用户，permissions 为权限，其他分别对应表名、列族及列名。

示例 10

收回用户 bigdata 对表 gamer_namespace:gamer 的 RWCA 权限。

关键代码：

```
hbase(main):007:0> revoke 'bigdata', 'RWCA','gamer_namespace:gamer'
0 row(s) in 0.4510 seconds
```

命令执行完以后，读者可以使用 user_permission 命令查看 gamer_namespace:gamer 表的权限来验证权限是否取消成功。

8.2.4　技能实训

完成对《英雄联盟》游戏玩家信息表的权限管理操作。

关键步骤：

（1）给用户 admin 授予 8.1.3 节技能实训中创建的 lol_gamer 表的 RWCA 权限。

（2）使用 USER_PERMISSION 命令查看 lol_gamer 表的权限。

（3）使用 REVOKE 命令收回 admin 用户对 lol_gamer 表的权限。

任务 3　理解 HRegion 切分

【任务描述】

理解 HRegion 切分的原理和策略。

【关键步骤】

（1）理解 HRegion 切分概念。

（2）了解 HRegion 切分策略。

（3）了解如何设置 HRegion 切分策略。

8.3.1　HRegion 切分概念

在前面讲解 HBase 的架构时曾介绍过，当 Region 达到一定的大小后会进行切分，一个 HRegion 会切分成两个大小相同的 HRegion，这样可以保证一个 HRegionServer 管理的 HRegion 大小不会过大，减轻了 HRegionServer 的压力。

在 HBase 中，HRegion 的大小设置是一个棘手的问题，需要综合考虑如下几个因素。

（1）HRegion 是 HBase 中分布式逻辑存储和负载均衡的最小单元，不同的 Region 会分布到不同的 RegionServer 上，但不是物理存储的最小单元。

（2）HRegion 由一个或者多个 HStore 组成，每个 HStore 保存一个列族，每个 HStore 由一个 MemStore 及 0 个或多个 Store File 组成。

（3）有时 HRegion 数目太多会造成性能下降。

（4）HRegion 数目太小会影响可扩展性，降低并行能力，有时还会导致压力不够分散。

HRegion 的切分操作是不可见的，因为 HMaster 不会参与其中。HRegionServer 拆分 HRegion 的步骤是，先将 HRegion 下线，然后切分，将切分后的子 HRegion 加入到.META.元数据信息表中，再将它们加入到 HRegionServer 中，最后汇报给 HMaster。在这个过程中，执行拆分的线程是 CompactSplitThread。

8.3.2　切分策略

1．切分策略

在 HBase 中，可以设置不同的 HRegion 切分策略来优化集群。

（1）ConstantSizeRegionSplitPolicy。在 Apache HBase 0.94 版本以前，该切分策略是默认而且唯一的切分策略。当 HRegion 中 HStore 的大小达到配置的 hbase.hregion.max.filesize（默认是 10GB）时触发切分操作。

（2）IncreasingToUpperBoundRegionSplitPolicy。在 Apache HBase 0.94 版本后，默认采用这种切分策略，该策略使用的最大 StoreFile 大小根据公式 Min(R^2*"hbase.hregion.memstore.flush.size","hbase.hregion.max.filesize")计算得出，其中，R 表示同一个 HRegionServer 上的 HRegion 的个数，hbase.hregion.memstore.flush.size 表示触发 MemStore 刷出缓存的 MemStore 的大小。比如，MemStoreu 阈值为 128MB，默认的 HRegion 最大值为 10GB，第一个 HRegion 将在 128MB 时切分。随着 HRegion 个数的增加，将会使用 512MB、1152MB、3200MB、6272MB 等作为切分点（Split Point），在达到 9 个 HRegion 之后，切分的大小会超过默认的 HRegion 最大值（10GB），之后便会使用 10GB 作为切分的大小。

（3）KeyPrefixRegionSplitPolicy。通过用户设置 Rowkey 的前缀来保证具有相同前缀的行都在一个 HRegion 中。指定 Rowkey 的前缀位数来划分 HRegion，需要读取表的 prefix_split_key_policy.prefix_length（前缀长度，数字类型），在切分时，按此长度对 SplitPoint 进行截取。这种策略比较适合固定前缀长度的 Rowkey，当没有指定前缀长度属性时，与使用 IncreasingToUpperBoundRegionSplitPolicy 策略的效果一样。

2. **配置切分策略**

HBase HRegion 既可以在 HBase 配置文件中定义全局切分策略，也可以在创建表和修改表的时候指定切分策略。用户可以根据不同的业务场景需要选择不同的切分策略。

（1）全局切分策略

指定全局切分策略是通过在 HBase 配置文件 hbase-site.xml 中配置 hbase.regionserver.region.split.policy 属性来实现的，具体配置如下。

```
<property>
  <name>hbase.regionserver.region.split.policy</name>
  <value>org.apache.hadoop.hbase.regionserver
           .IncreasingToUpperBoundRegionPolicy</value>
</property>
```

（2）创建表的时候指定切分策略

```
create 'table',{NAME => 'cf',SPLIT_POLICY =>
    'org.apache.hadoop.hbase.regionserver.ConstantSizeRegionSplitPolicy'}
```

从上面的命令可以看出，使用 SPLIT_POLICY 指定拆分策略。

HBase Region 切分原理

任务 4 了解 HBase 中的 Compaction 过程

【任务描述】

了解在 HBase 中 Compaction 的意义及实现方式。

【关键步骤】

（1）了解 HBase 中 Compaction 的概念。

（2）了解 HBase 中 Compaction 的实现方式。

8.4.1 Compaction 概念

在前面的章节中介绍过，HBase 中一个表的数据会存储到 HRegionServer 管理的一个 HRegion 上，HRegion 上的每一个 Column Family 会有一个 MemStore，当 MemStore 达到 hbase.hregion.memstore.flush.size 限制的值时，MemStore 会将它的数据内容刷出（Flush）到 StoreFile（HFile）中。当 StoreFile 的数量越来越多时，一次查询就需要很多的 IO 操作，这样会严重影响 HBase 的读性能。为了防止小文件过多，以保证查询效率，HBase 在必要的时候会将这些小文件合并成相对较大的 StoreFile，这个合并（merge）的过程就称为 Compaction。

Compaction 的作用主要包含以下三个。

（1）合并文件。

（2）清除已经删除、过期和多余版本的数据。

（3）提高读写数据的效率。

8.4.2 Compaction 实现方式

HBase 中提供了两种 Compaction 的实现方式，分别为 Minor 和 Major。

（1）Minor Compaction。通常选择几个临近的小的 StoreFile，把它们重写成一个大的 StoreFile。出于性能方面的考虑，Minor Compaction 不会删除过期的或者标记为要删除的数据，其结果是在一个 HStore 中生成更少、更大的 StoreFile。

（2）Major Compaction。对 HStore 下的所有 StoreFile 执行合并操作，其结果是为每个 HStore 生成一个 StoreFile。在生成的过程中，会删除已经标记为要删除的数据。

这两种方式都可以将 HBase 中的 StoreFile 进行合并，但是在 Compaction 操作期间，会影响 HBase 集群的性能，比如占用网络 IO、磁盘 IO 等，因此 Compaction 操作的实质就是在短时间内，通过消耗网络 IO 和磁盘 IO 等机器资源来换取 HBase 的读写性能。这就决定了只有在 HBase 集群空闲的时段才能做 Compaction 操作。

8.4.3 Compaction 参数

HBase Compaction 的触发因子可以在 HBase 的配置文件 hbase-site.xml 中进行配置。

Minor Compaction 的触发由以下几个参数共同决定。

（1）hbase.hstore.compaction.min。最小 Minor Compaction 的文件个数，默认值是 3，表示至少需要 3 个满足条件的 StoreFile，Minor Compaction 才会启动。

（2）hbase.hstore.compaction.max。最大 Minor Compaction 的文件个数，默认值是 10，表示一次 Minor Compaction 最多选择 10 个 StoreFile。

HBase Compaction 配置

（3）hbase.hstore.compaction.min.size。选取 Minor Compaction 文件时的参数值，表示文件大小小于该值的 StoreFile 一定会加入到 Minor Compaction 的 StoreFile 中。

（4）hbase.hstore.compaction.max.size。选取 Minor Compaction 文件时的参数值，表示文件大小大于该值的 StoreFile 一定会被 Minor Compaction 排除。

（5）hbase.hstore.compaction.ratio。将 StoreFile 按照年龄（old –>younger）进行排序，Minor Compaction 总是从 old StoreFile 开始选择，如果选择的 StoreFile 的大小小于它后面 hbase.hstorecompaction.max 个 StoreFile 大小之和乘以 ratio 的值，则该 StoreFile 也将加入到 Minor Compaction 中。

Major Compaction 的触发则相对简单。

（1）自动触发。配置 hbase.hregion.majorcompaction 参数，单位为毫秒。默认值是 604800000，表示一周执行一次 Major Compaction。

（2）手动触发。在 HBase Shell 中使用命令“major_compact 'tablename'”可以手动触发 Major Compaction。

本章总结

➢ HBase 名字空间的特性是对表资源进行隔离的一种技术，该隔离技术是决定 HBase 能否实现资源统一化管理的关键，提高了整体的安全性。

➢ 在实际工作中，可以通过设置对表或者名字空间的访问权限来保证 HBase 中数据的安全性。

➢ HRegion 切分策略的选择对 HBase 的性能有很大的影响，用户需根据业务场景选择合适的切分策略，这样可以提高 HBase 的性能。

➢ HBase 中支持 Minor 和 Major 两种 Compaction 来对 StoreFile 进行合并，能极大地提高 HBase 的读性能。

本章作业

一、简答题

1．HBase 中引入名字空间特性的好处是什么？

2．HBase 提供了哪些安全管控级别？

二、编码题

创建管理学生成绩表的名字空间 stu_scores_namespace，完成以下操作：

1．在 stu_scores_namespace 名字空间下创建学生成绩表 scores，包括列族 grade（年级）和 course（课程），学生姓名 name 作为行键。

2．给用户 teacher 授予 scores 表的 RWCA 权限。

3．查看名字空间 stu_scores_namespace 的详细信息。

4．查看 scores 表的权限信息。

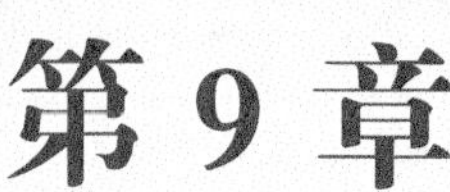

第9章 工作流调度框架 Oozie

技能目标

- 理解 Oozie 的架构及执行流程
- 学会搭建 Oozie 环境
- 学会在 Oozie 上进行作业调度

本章任务

任务1　理解 Apache Oozie 架构
任务2　搭建 Oozie 环境
任务3　实现游戏玩家搜索功能

本章资源下载

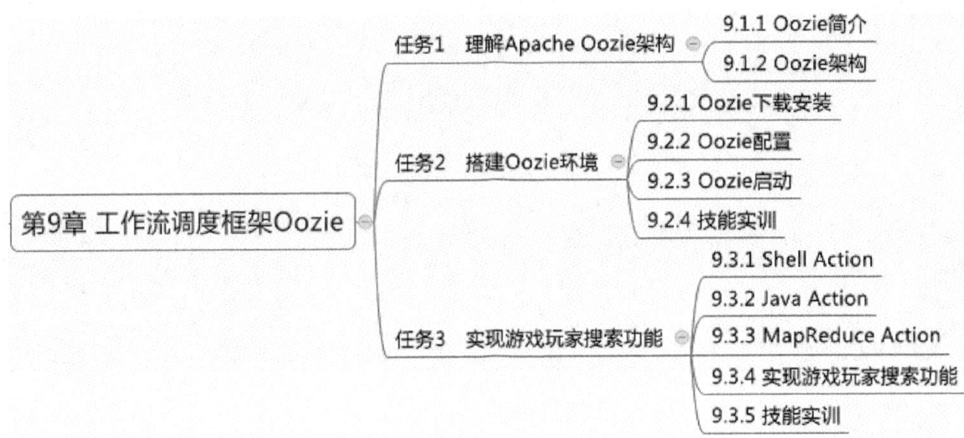

Oozie 是 Cloudera 公司共享给 Apache 的一个开源顶级项目，用于 Hadoop 平台的任务调度，是目前使用非常广泛的一种工作流调度引擎。本章主要介绍 Oozie 的概念及架构、Oozie 的作用、如何搭建 Oozie 环境，并在 Oozie 环境上完成几种 Action 的操作。

任务 1 理解 Apache Oozie 架构

【任务描述】

理解 Oozie 的概念及架构设计，了解 Oozie 的作用及应用场景。

【关键步骤】

（1）了解 Oozie 的概念、作用及应用场景。

（2）理解 Oozie 的架构及组件的功能职责。

9.1.1 Oozie 简介

Apache Oozie 是用于 Hadoop 平台的一种工作流调度引擎。在实际开发工作中，通常会遇到这样的场景，一个任务可能需要好几个 Hadoop 作业（Job）来协作完成，往往一个作业的输出会被当作另一个作业的输入来使用，这就涉及到了数据流的处理，即需要对作业进行统一的管理和调度。Oozie 是一个作业协调工具，它可以把多个 MapReduce 作业组合到一个逻辑工作单元中，从而完成更大型的任务。

9.1.2 Oozie 架构

1. *架构简介*

Oozie 是一种 Java Web 应用程序，它运行在 Java Servlet 容器（Tomcat）中，并且使用关系型数据库来存储以下内容。

（1）工作流定义。

（2）当前运行的工作流实例，包括实例的状态和变量。

Oozie 架构如图 9.1 所示。

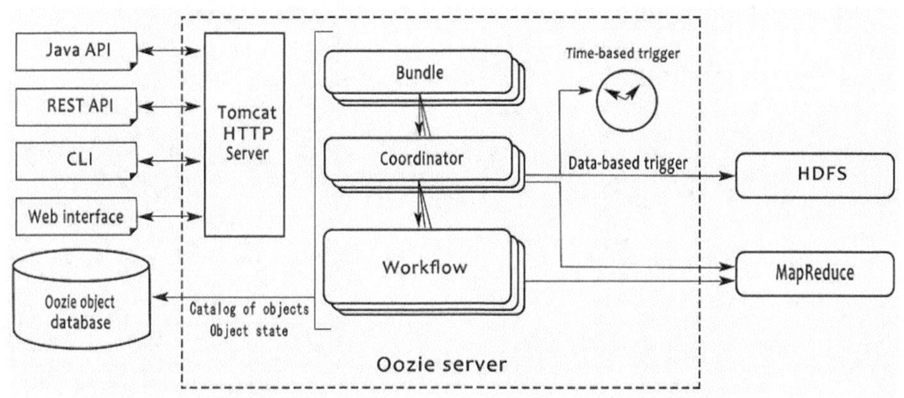

图9.1 Oozie架构图

从结构图中可以看出，Oozie 架构分为三部分。

左侧：Oozie 通过 Tomcat HTTP Server 对外提供了 Java API、REST API、CLI、Web 接口（Hue），产生的数据都存储在 Oozie Object Database 上。

中间：Oozie 的三层架构，包括 Bundle、Coordinator、Workflow。

右侧：Oozie 的 Coordinator Engine 协调引擎能够监控基于时间的触发器和基于 HDFS 数据的触发器。

2. Oozie Server 组件及组件之间的关系

从图 9.1 中可以看出，在 Oozie Server 中有三个重要的组件：Workflow、Coordinator、Bundle。

Workflow：工作流，是放置在控制依赖 DAG（有向无环图）中的一组动作（如 Hadoop 的 MapReduce 作业、Pig 作业等），其中指定了动作执行的顺序，使用 hPDL（一种 xml 流程定义语言）来描述 DAG。

Coordinator：协调器，可以理解为工作流的协调器，将多个工作流协调成一个工作流进行处理。

Bundle：捆，束，将一堆 Coodinator 进行汇总处理。

简单来说，Workflow 是对要进行的顺序化工作的抽象，Coordinator 是对要进行的顺序化的 Workflow 的抽象，Bundle 是对一堆 Coordinator 的抽象。三个组件之间的关系描述如下。

- 一个 Bundle 作业可以有一个或者多个 Coordinator 作业。
- 一个 Coordinator 作业可以有一个或者多个 Workflow 作业。

一个 Oozie 作业由三部分组成，分别是 job.properties、workflow.xml、lib 文件夹。

（1）job.properties

从名称可以看出，job.properties 是用来配置作业（Job）中用到的各种参数的，参数的具体描述如表 9-1 所示。

表 9-1 job.properties 参数

参数名称	参数含义
nameNode	hdfs 地址
jobTracker	jobTracker（ResourceManager）地址
queueName	Oozie 队列（默认填写 default）
examplesRoot	全局目录（默认填写 examples）
oozie.usr.system.libpath	是否加载用户 lib 目录
oozie.libpath	用户 lib 库所在的位置
oozie.wf.application.path	Oozie 流程所在 hdfs 地址（workflow.xml 地址）
user.name	当前用户
oozie.coord.application.path	coordinator.xml 地址（没有可以不写）
oozie.bundle.application.path	bundle.xml 地址（没有可以不写）

（2）workflow.xml

workflow.xml 是定义任务的整理流程的文件。定义了两种工作流节点，分别是流程控制节点（Control Flow Node）和动作节点（Action Node）。

① 流程控制节点

➢ start 节点

```
<workflow-app name= "[WF-DEF-NAME] " xmlns= " uri:oozie:workflow:0.4">
    …
    <start to= " [NODE-NAME]"/>
    …
</workflow-app>
```

其中，start 标签的 to 属性指向第一个要执行的工作流节点。

➢ end 节点

```
<workflow-app name= "[WF-DEF-NAME] " xmlns= " uri:oozie:workflow:0.4">
    …
    <end name= " [NODE-NAME]"/>
    …
</workflow-app>
```

到达该节点，工作流 Job 会变成 success 状态，表示成功完成。需要注意的是，一个工作流定义中必须且只能有一个 end 节点。

➢ kill 节点

```
<workflow-app name= "[WF-DEF-NAME] " xmlns= " uri:oozie:workflow:0.4">
    …
    <kill name= " [NODE-NAME]">
        <message>[MESSAGE-TO-LOG]</message>
    </kill>
    …
```

```
</workflow-app>
```

kill 元素的 name 属性是要杀死的工作流节点的名称，message 元素指定了工作流节点被杀死的备注信息。到达该节点，工作流 Job 会变成 KILLED 状态。

➢　decision 节点

```
<workflow-app name= "[WF-DEF-NAME] " xmlns= " uri:oozie:workflow:0.4">
    …
    <decision name= " [NODE-NAME]">
        <switch>
            <case to= " [NODE-NAME]">[PREDICATE]</case>
            …
            <case to= " [NODE-NAME]">[PREDICATE]</case>
            <default to= " [NODE-NAME]" />
        </switch>
    </decision>
    …
</workflow-app>
```

decision 节点类似于 Java 中的选择结构，通过预定义一组条件，当工作流 Job 执行到该节点时，会根据其中的条件进行判断，满足条件的路径被执行。decision 节点通过 switch…case 语法进行路径选择，只要有满足条件的判断，就会执行对应的路径；如果没有，则会执行 default 元素指向的节点。

➢　fork 节点和 join 节点

```
<workflow-app name= "[WF-DEF-NAME] " xmlns= " uri:oozie:workflow:0.4">
    …
    <fork name= "[FORK-NODE-NAME]">
        <path start="[NODE-NAME]" />
        …
        <path start="[NODE-NAME]" />
    </fork>
    <join name="[JOIN-NODE-NAME]" to="[NODE-NAME]" />
    …
</workflow-app>
```

fork 节点下会有多个 path 节点，指定了可以并发执行的多个执行路径。fork 中的多个并发路径会在 join 节点的位置汇合，当所有的 path 路径都到达后，才会继续执行 join 节点。fork 和 join 节点必须是成对出现的。

②　动作节点

在流程定义中，动作节点是能够触发一个计算任务或者处理任务执行的节点。所有的动作节点都具有一些基本的特性。

异步性（Asynchronous）：动作节点的执行，对于 Oozie 来说是异步的。Oozie 启动一个工作流 Job，这个工作流 Job 便开始执行。

可恢复性（Action Recovery）：一个动作节点执行失败，Oozie 会为其提供恢复策略。如果是状态转移过程中的失败，Oozie 会根据指定的重试时间间隔去重新执行；如果不是转移性质的失败，则只能通过手动干预进行恢复，如果恢复仍没有解决问题，最终会跳转到 kill 节点。

Oozie 内置支持的动作节点包括 MapReduce Action、FS（HDFS）Action、Java Action、Shell Action、Hive Action、Sqoop Action、DISCP Action、Spark Action 等。

任务 2 搭建 Oozie 环境

【任务描述】

学会搭建 Oozie 环境。

【关键步骤】

（1）下载 Oozie 安装包并解压安装。

（2）配置 Oozie。

（3）启动及验证。

9.2.1 Oozie 下载安装

1. 安装包下载

本书中 Oozie 采用的是 CDH 版本的 Oozie-4.1.0-cdh5.14.2，而且是 CDH 已经编译完成的版本，读者可以去 Cloudera 官网下载对应的版本。软件包下载完成后，可以通过远程工具上传到自己的虚拟机服务器上，如 /home/hadoop/software 目录下。Oozie 环境安装步骤请扫描二维码。

Oozie 环境搭建

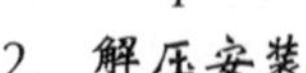

2. 解压安装

（1）下载完成以后，需要将安装包解压。

```
[hadoop@hadoop ~]$ tar –zxvf /home/hadoop/software/oozie-4.1.0-cdh5.14.2.tar.gz
```

（2）将解压后的文件复制到/opt 目录下。

```
[hadoop@hadoop ~]$ sudo mv /home/hadoop/software/oozie-4.1.0-cdh5.14.2
/opt/oozie-4.1.0-cdh5.14.2
```

（3）配置环境变量。

```
[hadoop@hadoop ~]$ vi ~/.bashrc
```

在打开的文件中添加如下两行：

```
export OOZIE_HOME=/opt/oozie-4.1.0-cdh5.14.2
export PATH=$OOZIE_HOME/bin:$PATH
```

使配置文件生效：

```
[hadoop@hadoop ~]$ source ~/.bashrc
```

注意

在安装 Oozie 之前，需要确保 JDK 和 Hadoop 环境已经搭建完成。本书选择使用外部 MySQL 存储 Oozie 元数据，还需要读者自己完成 MySQL 环境的搭建。

9.2.2 Oozie 配置

Oozie 解压安装并配置好环境变量后，前置工作就完成了，接下来就要对 Oozie 进行配置。

1. **配置 core-site.xml 文件**

由于 Oozie 需要访问 Hadoop 文件系统，必须先在 core-site.xml 文件中配置 Oozie 程序运行的用户，以免没有访问权限。该配置文件在$HADOOP_HOME/etc/hadoop 目录下，配置内容如下。

```
<!-- OOZIE -->
<property>
    <name>hadoop.proxyuser.[OOZIE_SERVER_USER].hosts</name>
    <value>[OOZIE_SERVER_HOSTNAME]</value>
</property>
<property>
    <name>hadoop.proxyuser.[OOZIE_SERVER_USER].groups</name>
    <value>[USER_GROUPS_THAT_ALLOW_IMPERSONATION]</value>
</property>
```

其中，[OOZIE_SERVER_USER] 填写访问 Tomcat 的用户名，[OOZIE_SERVER_HOSTNAME]填写安装 Tomcat 服务器的机器的主机名，[USER_GROUPS_THAT_ALLOW_IMPERSONATION]一般填写“*”标识符，表示所有用户都可以访问。

配置 core-site.xml。

```
<!-- OOZIE -->
<property>
    <name>hadoop.proxyuser.hadoop.hosts</name>
    <value>hadoop</value>
</property>
<property>
    <name>hadoop.proxyuser.hadoop.groups</name>
    <value>*</value>
</property>
```

2. **配置 oozie-site.xml 文件**

Oozie 的配置文件在$OOZIE_HOME/conf/目录下，在 oozie-site.xml 配置文件中添加如下代码。

```
<!--设置 Hadoop 的配置文件的路径-->
<property>
    <name>oozie.service.HadoopAccessorService.hadoop.configurations</name>
    <value>*=/opt/hadoop-2.6.0-cdh5.14.2/etc/hadoop</value>
</property>
<!--配置 MySQL，作为 Oozie 元数据存放的数据库 -->
<property>
    <name>oozie.service.JPAService.jdbc.driver</name>
```

9 Chapter

```
        <value>com.mysql.jdbc.Driver</value>
        <description>JDBC driver class</description>
</property>
<property>
        <name>oozie.service.JPAService.jdbc.url</name>
        <value>jdbc:mysql://hadoop:3306/oozie?createDatabaseIfNotExist=true</value>
        <description>JDBC URL</description>
</property>
<property>
        <name>oozie.service.JPAService.jdbc.username</name>
        <value>root</value>
        <description>DB user</description>
</property>
<property>
        <name>oozie.service.JPAService.jdbc.password</name>
        <value>root</value>
        <description>DB user password</description>
</property>
```

配置完 core-site.xml 文件后，需要重启 Hadoop 集群，让配置文件生效。

3. 配置 Oozie Server

配置文件配置好以后，就需要配置 Oozie Server，具体步骤如下。

（1）在$OOZIE_HOME/目录下创建 libext 目录。

```
[hadoop@hadoop ~]$ cd /opt/oozie-4.1.0-cdh5.14.2
[hadoop@hadoop oozie-4.1.0-cdh5.14.2]$ mkdir libext
```

（2）解压 hadooplibs.tar.gz。

① Oozie 解压完成后，在$OOZIE_HOME/目录下会存在 hadooplibs 的压缩包 oozie-hadooplibs-4.0.0-cdh5.3.6.tar.gz，解压该文件。

```
[hadoop@hadoop oozie-4.1.0-cdh5.14.2]$ sudo tar -zxvf oozie-hadooplibs-4.1.0-cdh5.14.2.tar.gz
```

② 解压完成后，在$OOZIE_HOME/目录下会存在 oozie-4.1.0-cdh5.14.2 文件夹，该文件夹下存在 hadooplibs 目录。

```
[hadoop@hadoop oozie-4.1.0-cdh5.14.2]$ cd $OOZIE_HOME/oozie-4.1.0-cdh5.14.2
[hadoop@hadoop oozie-4.1.0-cdh5.14.2]$ ll
总用量 4
drwxr-xr-x. 4 hadoop hadoop 4096 3 月 28 04:31 hadooplibs
```

③ 切换到 hadooplibs 目录下，存在支持 MapReduce1.x 和 YARN 的 jar 目录。

```
[hadoop@hadoop oozie-4.1.0-cdh5.14.2]$ cd $OOZIE_HOME/oozie-4.1.0-cdh5.14.2/hadooplibs
[hadoop@hadoop hadooplibs]$ ll
总用量 8
```

```
drwxr-xr-x. 2 hadoop hadoop 4096 3 月  28 04:31 hadooplib-2.6.0-cdh5.14.2.oozie-4.1.0-cdh5.14.2
drwxr-xr-x. 2 hadoop hadoop 4096 3 月  28 04:31 hadooplib-2.6.0-mr1-cdh5.14.2.oozie-4.1.0-cdh5.14.2
```

（3）将 hadooplibs 下支持 YARN 的 jar 文件拷贝到创建的$OOZIE_HOME/libext 目录下。

```
[hadoop@hadoop ~]$ cd $OOZIE_HOME/oozie-4.1.0-cdh5.14.2/hadooplibs/
[hadoop@hadoop ~]$ cp hadooplib-2.6.0-cdh5.14.2.oozie-4.1.0-cdh5.14.2/* $OOZIE_HOME/libext
```

（4）下载 ExtJs2.2。由于 Oozie 前端页面是用 ExtJS 编写的，所以需要将 ExtJS2.2 的包放到$OOZIE_HOME/libext 目录下。读者可以去 Cloudera 官网下载 ExtJS2.2 的包，并把下载的包放到$OOZIE_HOME/libext 目录下。

（5）由于使用 MySQL 作为 Oozie 元数据的数据库，所以需要将 MySQL 驱动包复制到$OOZIE_HOME/libext 目录下以及$OOZIE_HOME/lib 目录下。

（6）执行 bin/oozie-setup.sh prepare-war 命令生成 war 包。Oozie 可执行命令都放在$OOIZE_HOME/bin 目录下。

```
[hadoop@hadoop ~]$ cd $OOZIE_HOME/
[hadoop@hadoop oozie-4.1.0-cdh5.14.2]$ bin/oozie-setup.sh prepare-war
```

命令执行完成后，在$OOZIE_HOME/oozie-server/webapps 目录下会生成 oozie.war。

```
[hadoop@hadoop ~]$ cd $OOZIE_HOME/oozie-server/webapps
[hadoop@hadoop webapps]$ ll
-rw-rw-r--.  1   hadoop   hadoop   197047179   7 月   13   11:37   oozie.war
drwxr-xr-x   3   hadoop   hadoop          55   3 月   28   04:40   ROOT
```

其中，$OOZIE_HOME/oozie-server/目录就是类似 Tomcat 的目录。

（7）执行如下命令创建共享包。

```
[hadoop@hadoop ~]$ $OOZIE_HOME/bin/oozie-setup.sh sharelib create -fs hdfs://192.168.85.239:
8020 -locallib $OOZIE_HOME/oozie-sharelib-4.1.0-cdh5.14.2-yarn.tar.gz
```

其中，hdfs://192.168.85.239:8020 是 HDFS 的访问地址。执行完该命令后，在 HDFS 文件系统中对应的用户目录下会创建 share 目录。

```
[hadoop@hadoop ~]$ hdfs dfs -ls /user/hadoop/
drwxr-xr-x    - hadoop   supergroup    0   2018-07-17 13:38 /user/hadoop/share
```

（8）执行如下命令创建数据库。

```
[hadoop@hadoop bin]$ $OOZIE_HOME/bin/ooziedb.sh create -sqlfile oozie.sql -run DB Connection
```

执行成功后，在$OOZIE_HOME/目录下会生成 oozie.sql 文件。

注意

在执行创建数据库命令之前，需保证 MySQL 数据库服务已经启动，并且保证 oozie-site.xml 文件中关于 MySQL 的配置属性正确，同时在$OOZIE_HOME/libext 和$OOZIE_HOME/lib 目录中已经放置有 MySQL 的驱动包。

执行完上述八个步骤后，Oozie Server 的配置就完成了。

9.2.3 Oozie 启动

Oozie 配置文件修改完成以及 Oozie Server 相关配置完成以后，就需要启动 Oozie。

1. *启动*

启动和关闭 Oozie 服务的相关命令都在$OOZIE_HOME/bin 目录下。

启动服务命令如下。

```
[hadoop@hadoop ~]$ $OOZIE_HOME/bin/oozied.sh start
```

停止服务命令如下。

```
[hadoop@hadoop ~]$ $OOZIE_HOME/bin/oozied.sh stop
```

2. **验证**

Oozie 启动完成后，有以下两种验证方式。

（1）使用 Java 命令 jps 查看进程。Oozie 启动后，进程列表中会存在 Bootstrap 进程。

```
[hadoop@hadoop bin]$ jps
26481 NameNode
26769 SecondaryNameNode
26913 ResourceManager
32449 Bootstrap
27349 QuorumPeerMain
27014 NodeManager
32534 Jps
26602 DataNode
```

从进程列表中可以看出存在 Bootstrap 进程，证明 Oozie 服务已经启动。

（2）使用 webui 的方式验证。Oozie 是 Tomcat 服务，提供了 webui 的操作方式，Oozie 启动后，可以使用网址 http://hadoop:11000/oozie/来访问。

其中，hadoop 是启动 Ooize 服务的机器的主机名。如果网页可以正常访问，证明 Oozie 服务已经成功启动。

9.2.4 技能实训

按照任务 2 的步骤，搭建自己的 Oozie 环境。

关键步骤：

（1）下载对应版本的安装包，安装并配置环境变量。

（2）配置 Oozie（配置文件的修改及 Oozie Server 的配置）。

（3）启动 Oozie 并验证是否启动成功。

任务 3 实现游戏玩家搜索功能

【任务描述】

使用 Oozie 工作流实现游戏玩家搜索功能。

【关键步骤】

（1）Shell Action 操作。

（2）Java Action 操作。

（3）MapReduce Action 操作。

（4）Oozie Scheduling 操作。

9.3.1 Shell Action

任务 1 已经介绍过动作节点，Oozie 中内置了 Shell Action。Shell Action 可以执行 Shell 命令，命令所需参数通过配置完成。Oozie Action Node 详细操作请扫描二维码。

Oozie Action Node 配置

示例 1

运行 Shell 脚本接收传递参数，根据参数名称在 hdfs 上创建对应目录。

关键步骤：

（1）创建本地的测试节点文件夹。

```
[hadoop@hadoop ~]$ mkdir -p ~/ooize/shell
```

（2）在创建的文件夹下新建 Shell 脚本 script.sh，并添加如下内容。

```
#!/bin/bash
dir_name=/home/hadoop/oozie/apps/shell/shell_script/$1
hdfs dfs –mkdir –p $dir_name
```

（3）在创建的文件夹下新建 job.properties 文件，并添加如下内容。

```
nameNode=hdfs://hadoop:8020
jobTracker=hadoop:8032
queueName=default
examplesRoot=oozie
oozie.wf.application.path=${nameNode}/home/${user.name}/${examplesRoot}/apps/shell/shell_script/workflow.xml
EXEC=script.sh   #步骤 2 创建的脚本文件
```

（4）在创建的文件夹下新建 workflow.xml 文件，并添加如下内容。

```
<workflow-app xmlns="uri:oozie:workflow:0.4" name="shell-wf">
    <start to="shell-node"/>
    <action name="shell-node">
        <shell xmlns="uri:oozie:shell-action:0.2">
            <job-tracker>${jobTracker}</job-tracker>
            <name-node>${nameNode}</name-node>
            <configuration>
                <property>
                    <name>mapred.job.queue.name</name>
                    <value>${queueName}</value>
                </property>
            </configuration>
            <exec>${EXEC}</exec>
```

```
                <!-- 创建 A 目录   -->
                <argument>A</argument>
                <file>${EXEC}#${EXEC}</file>
            </shell>
            <ok to="end"/>
            <error to="fail"/>
        </action>
        <kill name="fail">
            <message>
                Shell action failed, error message[${wf:errorMessage(wf:lastErrorNode())}]
            </message>
        </kill>
        <end name="end"/>
</workflow-app>
```

<file>元素会复制指定的文件到运行该脚本的机器上。

（5）在 HDFS 上创建目录。

```
[hadoop@hadoop ~]$ hdfs dfs -mkdir -p /home/hadoop/oozie/apps/shell/shell_script/
```

（6）上传 script.sh、workflow.xml 到 HDFS。

```
[hadoop@hadoop ~]$ hdfs dfs -put /home/hadoop/ooize/shell/workflow.xml /home/hadoop/oozie/apps/
shell/shell_script/
[hadoop@hadoop ~]$ hdfs dfs -put ~/ooize/shell/script.sh /home/hadoop/oozie/apps/shell/shell_script/
```

（7）使用 Oozie 命令执行任务。

```
[hadoop@hadoop ~]$ oozie job -oozie http://hadoop:11000/oozie -config
/home/hadoop/oozie/shell/job.properties –run
Job ID : 0000000-180719134309717-oozie-hado-W
```

（8）查看作业信息。

```
[hadoop@hadoop ~]$ oozie job -oozie http://hadoop:11000/oozie -info 0000000-180719134309717-
oozie-hado-W
Job ID : 0000000-180719134309717-oozie-hado-W
------------------------------------------------------------------------------------------------------------
Workflow Name : shell-wf
App Path      : hdfs://hadoop:8020/home/hadoop/oozie/apps/shell/shell_script/workflow.xml
Status        : SUCCEEDED
Run           : 0
User          : hadoop
Group         : -
Created       : 2018-07-19 09:41 GMT
Started       : 2018-07-19 09:41 GMT
Last Modified : 2018-07-19 09:41 GMT
Ended         : 2018-07-19 09:41 GMT
CoordAction ID : -

Actions
------------------------------------------------------------------------------------------------------------
```

```
ID                                                      Status    Ext ID    Ext Status    Err Code
------------------------------------------------------------------------------------------------------------
0000014-180719134309717-oozie-hado-W@:start:             OK        -         OK            -
------------------------------------------------------------------------------------------------------------
0000014-180719134309717-oozie-hado-W@shell-node OK  job_1531978002849_0028 SUCCEEDED
-

------------------------------------------------------------------------------------------------------------
0000014-180719134309717-oozie-hado-W@end                 OK        -         OK            -
------------------------------------------------------------------------------------------------------------
```

（9）在 HDFS 上查看是否创建成功。

```
[hadoop@hadoop ~]$ hdfs dfs -ls /home/hadoop/oozie/apps/shell/shell_script
Found 3 items
drwxr-xr-x  -   hadoop  supergroup    0   2018-07-19  17:41  /home/hadoop/oozie/apps/shell/ shell_
script/A
-rw-r--r--    1   hadoop supergroup 1003 2018-07-19 17:39
/home/hadoop/oozie/apps/shell/shell_script/script.sh
-rw-r--r--    1   hadoop supergroup 1645 2018-07-19 17:39
/home/hadoop/oozie/apps/shell/shell_script/workflow.xml
```

从结果中可以看出，在 HDFS 上，/home/hadoop/oozie/apps/shell/shell_script/A 目录已经成功创建。

9.3.2 Java Action

Oozie 提供了对 Java Action 的支持，Java Action 用于执行一个具有 main 方法的应用程序。在 Oozie 工作流定义中，Java Action 会作为一个 MapReduce 作业执行，这个作业只有一个 Map 任务，该任务需要在 job.properties 中指定 nameNode、jobTracker 的信息，并配置 Java 应用程序的 JVM 选项参数，以及给主函数传递参数。

示例 2

Java Action 配置。

关键步骤：

（1）创建本地的测试节点文件夹。

```
[hadoop@hadoop ~]$ mkdir -p /home/hadoop/ooize/java
```

（2）新建 job.properties 文件，并添加如下内容。

```
nameNode=hdfs://hadoop:8020
jobTracker=hadoop:8032
queueName=default
examplesRoot=oozie
oozie.wf.application.path=${nameNode}/home/${user.name}/${examplesRoot}/apps/java-main/
                            workflow.xml
```

（3）新建 workflow.xml 文件。

```
<workflow-app xmlns="uri:oozie:workflow:0.4" name="java-main-wf">
        <start to="java-node"/>
```

```
        <action name="java-node">
            <java>
                <job-tracker>${jobTracker}</job-tracker>
                <name-node>${nameNode}</name-node>
                <configuration>
                    <property>
                        <name>mapred.job.queue.name</name>
                        <value>${queueName}</value>
                    </property>
                </configuration>
                <main-class>org.apache.oozie.example.DemoJavaMain</main-class>
                <arg>Hello</arg>
                <arg>Oozie!</arg>
            </java>
            <ok to="end"/>
            <error to="fail"/>
        </action>
        <kill name="fail">
            <message>Java failed, error message[${wf:errorMessage(wf:lastErrorNode())}] </message>
        </kill>
        <end name="end"/>
    </workflow-app>
```

（4）复制 oozie examples 下的 jar 包到/home/hadoop/oozie/java。

```
[hadoop@hadoop ~]$ cp /opt/oozie-4.1.0-cdh5.14.2/examples/apps/java-main/lib /home/hadoop/oozie/java
```

（5）在 HDFS 上创建目录。

```
[hadoop@hadoop ~]$ hdfs dfs -mkdir -p /home/hadoop/oozie/apps/java-main
```

（6）上传 workflow.xml、lib 包到 HDFS 对应目录。

```
[hadoop@hadoop ~]$ hdfs dfs -put /home/hadoop/ooize/java/* /home/hadoop/oozie/apps/java-main/
```

（7）执行 oozie 命令。

```
[hadoop@hadoop ~]$ oozie job -oozie http://hadoop:11000/oozie -config /home/hadoop/oozie/java/job.properties –run
Job ID : 0000000-180719134309717-oozie-hado-W
```

（8）查看作业信息。

```
[hadoop@hadoop lib]$ oozie job -oozie http://hadoop:11000/oozie -info 0000000-180719134309717-oozie-hado-W
Job ID : 0000000-180719134309717-oozie-hado-W
------------------------------------------------------------------------------------------------------------------------
Workflow Name : java-main-wf
App Path      : hdfs://hadoop:8020/home/hadoop/oozie/apps/java-main
Status        : SUCCEEDED
Run           : 0
User          : hadoop
Group         : -
```

```
Created         : 2018-07-20 08:27 GMT
Started         : 2018-07-20 08:27 GMT
Last Modified   : 2018-07-20 08:28 GMT
Ended           : 2018-07-20 08:28 GMT
CoordAction ID : -

Actions
------------------------------------------------------------------------------------------------
ID                                                Status   Ext ID      Ext Status     Err Code
------------------------------------------------------------------------------------------------
0000000-180719134309717-oozie-hado-W@:start:      OK       -           OK             -
------------------------------------------------------------------------------------------------
0000000-180719134309717-oozie-hado-W@java-node    OK   job_1531978002849_0038 SUCCEEDED -
------------------------------------------------------------------------------------------------
0000000-180719134309717-oozie-hado-W@end          OK       -           OK             -
------------------------------------------------------------------------------------------------
```

9.3.3 MapReduce Action

Oozie 支持 MapReduce Action，MapReduce Action 会在工作流 Job 中启动一个 MapReduce Job 任务运行，需要在 job.properties 和 workflow.xml 文件中进行配置。

示例 3

Oozie 运行 MapReduce 作业。

分析：

Oozie 安装包提供了 MapReduce 测试案例，在$OOZIE_HOME 目录下存在压缩包 oozie-examples.tar.gz，该压缩包里存放了 Oozie 的各种 Action 的测试案例。

关键步骤：

（1）解压。

```
[hadoop@hadoop ~]$ cd /opt/oozie-4.1.0-cdh5.14.2
```

① 解压 oozie-examples.tar.gz

```
[hadoop@hadoop oozie-4.1.0-cdh5.14.2]$ tar -zxf oozie-examples.tar.gz
```

解压后会在当前目录下生成 examples 文件夹，接下来所要运行的 oozie 的测试案例都在该文件夹中。案例的运行需要一些库文件，这些库文件都在$OOZIE_HOME/oozie-sharelib-4.1.0-cdh5.14.2.tar.gz 压缩文件中。

② 解压 oozie-sharelib-4.1.0-cdh5.14.2.tar.gz

```
[hadoop@hadoop oozie-4.1.0-cdh5.14.2]$ tar -zxf oozie-sharelib-4.1.0-cdh5.14.2.tar.gz
```

解压后会在当前目录下生成 share 目录，里面的 jar 包是 oozie 运行时必备的库文件。

（2）将 examples 和 share 目录下的文件上传到 hdfs 中。

```
[hadoop@hadoop ~]$ hdfs dfs -put $OOZIE_HOME/examples    /user/hadoop/examples
[hadoop@hadoop ~]$ hdfs dfs -put $OOZIE_HOME/share    /user/hadoop/share
```

（3）修改本地配置文件 job.properties。

```
namenode=hdfs://hadoop:8020
```

```
jobTracker=hadoop:8032
queueName=default
examplesRoot=examples
oozie.wf.application.path=
        ${nameNode}/user/${user.name}${ examplesRoot }/apps/map-reduce/workflow.xml
outputDir=map-reduce
```

（4）执行 MapReduce 任务。

```
[hadoop@hadoop bin]$ oozie job -oozie http://hadoop:11000/oozie -config /opt/oozie-4.1.0-cdh5.14.2/
examples/apps/map-reduce/job.properties -run
Job ID : 0000000-180714095143226-oozie-hado-W
```

（5）查看执行情况。

```
[hadoop@hadoop ~]$ oozie job -oozie http://hadoop:11000/oozie -info 0000000-180714095143226-
oozie-hado-W
Job ID : 0000000-180719134309717-oozie-hado-W
------------------------------------------------------------------------------------------------
Workflow Name : map-reduce-wf
App Path      : hdfs://hadoop:8020/user/hadoop/examples/apps/map-reduce/workflow.xml
Status        : SUCCEEDED
Run           : 0
User          : hadoop
Group         : -
Created       : 2018-07-14 09:41 GMT
arted         : 2018-07-14 09:41 GMT
▽ast Modified : 2018-07-14 09:41 GMT
Ended         : 2018-07-14 09:41 GMT
CoordAction ID : -

Actions
------------------------------------------------------------------------------------------------
ID                                          Status   Ext ID      Ext Status     Err Code
------------------------------------------------------------------------------------------------
0000014-180719134309717-oozie-hado-W@:start:   OK       -           OK             -
------------------------------------------------------------------------------------------------
0000014-180719134309717-oozie-hado-W@shell-node  OK   job_1531978002849_0021  SUCCEEDED
-
------------------------------------------------------------------------------------------------
0000014-180719134309717-oozie-hado-W@end       OK       -           OK             -
------------------------------------------------------------------------------------------------
```

（6）查看 map-reduce 任务执行的结果。

```
[hadoop@hadoop ~]$ hdfs dfs -cat /user/hadoop/examples/output-data/map-reduce/part-00000
    0   To be or not to be, that is the question;
    42  Whether 'tis nobler in the mind to suffer
    84  The slings and arrows of outrageous fortune,
    129 Or to take arms against a sea of troubles,
```

```
172  And by opposing, end them. To die, to sleep;
217  No more; and by a sleep to say we end
255  The heart-ache and the thousand natural shocks
302  That flesh is heir to ? 'tis a consummation
```

9.3.4　实现游戏玩家搜索功能

前面介绍了 Oozie 的几种 Action 的配置以及如何运行，本节通过编写 MapReduce 作业来实现从各个大区中模糊搜索匹配的玩家游戏 ID，并提交到 Oozie 中运行。

示例 4

通过 Oozie 工作流运行游戏玩家 ID 搜索作业。

示例描述：

（1）三个 txt 格式的测试数据如下。

location1.txt 内容：

你的仓鼠太爱我。偷猪贼。吹牛大王。浅念。双木夕。星曜二。光曜二。会飞的猪。广东省第一蔡文姬。

一念。国士无双。三颗柚。了解。杨、鹏程。brown。inero。warshy。徐子陵。

钻石五。望穿秋水。保安科杨锦荣。老帅。蓝烟。张大仙。梦泪。二把刀。

location2.txt 内容：

杰斯。辰鬼。飞牛。小渝。流苏。猫神。阿泰。无痕。揪心。陌离。

恶女。魅魇。分岛花音。状况外。北岛。别荡失太早。天堂。今夜。

一个帅二个呆。幻听。二当家。温柔女 Boss。诱惑。查理 boy。病态二公主。

location3.txt 文件内容：

我的阳光。痕迹。最初的借口。说谎。沉迷者。冠军。疲惫。毕业生。青春呐喊。

凉城。只有你。影子爱人。小晴天。男友力。二叉树。黑人。断夏。小情绪。

叫我二宝贝。虚假。生命力。刺痛。二营长。醉墨卷秋澜。雪忆谰城。

（2）实现功能：查找名称包含“二”的所有玩家。

（3）预期结果：

```
location1.txt    二把刀---|---光曜二---|---星曜二---|---
location2.txt    病态二公主---|---二当家---|---一个帅二个呆---|---
location3.txt    二营长---|---叫我二宝贝---|---二叉树---|---
```

关键步骤：

（1）编写 MapReduce 程序，实现游戏玩家 ID 搜索功能。

关键代码：

① Mapper 类

```
package org.apache.oozie;
import org.apache.hadoop.io.Text;
import org.apache.hadoop.mapreduce.Mapper;
import org.apache.hadoop.mapreduce.lib.input.FileSplit;
```

```
import java.io.IOException;
import java.util.StringTokenizer;
public class Map extends Mapper<Object, Text, Text, Text> {
    private static final String word = "二";
    private FileSplit fileSplit;
    public void map(Object key,Text value,Context context) throws IOException,
        InterruptedException{
        fileSplit = (FileSplit)context.getInputSplit();
        String fileName = fileSplit.getPath().getName().toString();
        //按句号分割
        StringTokenizer st = new StringTokenizer(value.toString(),"。");
        while(st.hasMoreTokens()){
            String line = st.nextToken().toString();
            if(line.indexOf(word)>=0){
                context.write(new Text(fileName),new Text(line));
            }
        }
    }
}
```

② Reducer 类

```
package org.apache.oozie;
import org.apache.hadoop.io.Text;
import org.apache.hadoop.mapreduce.Reducer;
import java.io.IOException;
public class Reduce extends Reducer<Text,Text,Text,Text> {
    public void reduce(Text key,Iterable<Text> values,Context context) throws IOException,
    InterruptedException{
        String lines = "";
        for(Text value:values){
        lines += value.toString()+"---|---";
        }
        context.write(key, new Text(lines));
    }
}
```

③ Driver 类

```
package org.apache.oozie;
import java.io.IOException;
import org.apache.hadoop.conf.Configuration;
import org.apache.hadoop.fs.Path;
import org.apache.hadoop.io.Text;
import org.apache.hadoop.mapreduce.Job;
import org.apache.hadoop.fs.FileSystem;
import org.apache.hadoop.mapreduce.lib.input.FileInputFormat;
import org.apache.hadoop.mapreduce.lib.output.FileOutputFormat;
```

```
import org.apache.hadoop.util.GenericOptionsParser;
/**
* MapReduce 实现游戏玩家搜索
*/
public class PlayerSearch {
    public static void main(String[] args) throws IOException, ClassNotFoundException,
        InterruptedException {
        Configuration conf = new Configuration();
        args = new String[]{"hdfs://hadoop:8020/user/hadoop/oozie/input/search_in",
                    "hdfs://hadoop:8020/user/hadoop/oozie/output/search_out"};
        //检查运行命令
        String[] otherArgs = new GenericOptionsParser(conf,args).getRemainingArgs();
        if(otherArgs.length != 2){
            System.err.println("Usage search <int> <out>");
            System.exit(2);
        }
        //配置作业名
        Job job = Job.getInstance(conf);
        //配置各个作业类
        job.setJarByClass(PlayerSearch.class);
        job.setMapperClass(Map.class);
        job.setReducerClass(Reduce.class);
        job.setOutputKeyClass(Text.class);
        job.setOutputValueClass(Text.class);
        FileInputFormat.addInputPath(job, new Path(otherArgs[0]));
        FileSystem fs = FileSystem.get(conf);
        Path path = new Path(otherArgs[1]);
        if (fs.isDirectory(path)) {
            fs.delete(path, true);
        }
        FileOutputFormat.setOutputPath(job, new Path(otherArgs[1]));
        System.exit(job.waitForCompletion(true) ? 0 : 1);
    }
}
```

（2）提交作业运行。

① 创建本地的测试节点文件夹。

```
[hadoop@hadoop ~]$ mkdir -p /home/hadoop/ooize/apps/map-reduce
```

② 在步骤①创建的文件夹下新建 lib 目录并将写好的 MapReduce 任务打成 jar 包，稍后上传到/user/hadoop/ooize/apps/map-reduce/lib 目录下。

③ 新建 job.properties 文件，并添加如下内容。

```
nameNode=hdfs://hadoop:8020
jobTracker=hadoop:8032
queueName=default
examplesRoot=oozie
```

```
oozie.wf.application.path=
        ${nameNode}/user/${user.name}/${examplesRoot}/apps/map-reduce/workflow.xml
outputDir=search_out
```

④ 新建 workflow.xml 文件，并添加如下代码。

```
<workflow-app xmlns="uri:oozie:workflow:0.4" name="map-reduce-wf">
  <start to="mr-node"/>
  <action name="mr-node">
    <map-reduce>
      <job-tracker>${jobTracker}</job-tracker>
      <name-node>${nameNode}</name-node>
      <prepare>
        <delete path="${nameNode}/user/hadoop/${examplesRoot}/output/${outputDir}"/>
      </prepare>
      <configuration>
          <property>
              <name>mapred.job.queue.name</name>
              <value>${queueName}</value>
          </property>
          <property>
              <name>mapred.mapper.new-api</name>
              <value>true</value>
          </property>
          <property>
              <name>mapred.reducer.new-api</name>
              <value>true</value>
          </property>
          <property>
              <!-- 配置编写的 MapReduce 的 Mapper 类-->
              <name>mapreduce.job.map.class</name>
              <value>org.apache.oozie.Map</value>
          </property>
          <property>
              <!-- 配置编写的 MapReduce 的 Reducer 类-->
              <name>mapreduce.job.reduce.class</name>
              <value>org.apache.oozie.Reduce</value>
          </property>
          <property>
              <name>mapred.map.tasks</name>
              <value>1</value>
          </property>
          <property>
              <name>mapred.input.dir</name>
              <value>/user/hadoop/${examplesRoot}/input/search_in</value>
              </property>
          <property>
```

```
                <name>mapred.output.dir</name>
                <value>/user/hadoop/${examplesRoot}/output/${outputDir}</value>
            </property>
        </configuration>
    </map-reduce>
    <ok to="end"/>
    <error to="fail"/>
  </action>
  <kill name="fail">
    <message>
            Map/Reduce failed, error message[${wf:errorMessage(wf:lastErrorNode())}]
    </message>
  </kill>
  <end name="end"/>
</workflow-app>
```

⑤ 上传 oozie 目录下的文件并放入到相应的 hdfs 中的目录下。

```
[hadoop@hadoop ~]$ hdfs dfs -mkdir -p /user/hadoop/oozie
[hadoop@hadoop ~]$ hdfs dfs -put /home/hadoop/oozie    /user/hadoop/oozie
```

⑥ 在 hdfs 上创建输入路径，并将测试数据上传到新建的输入目录下。

```
[hadoop@hadoop ~]$ hdfs dfs -mkdir -p /user/hadoop/oozie/input/search_in
[hadoop@hadoop ~]$ hdfs dfs -put /home/hadoop/testData/location1.txt location2.txt location3.txt /user/hadoop/ooize/input/search_in/
```

⑦ 执行命令，执行作业。

```
[hadoop@hadoop bin]$ oozie job -oozie http://hadoop:11000/oozie -config /home/hadoop/oozie/apps/map-reduce/job.properties -run
Job ID : 0000000-180717091145523-oozie-hado-W
```

⑧ 查看作业信息。

```
[hadoop@hadoop bin]$ oozie job -oozie http://hadoop:11000/oozie -info 0000000-180717091145523-oozie-hado-W
Job ID : 0000000-180717091145523-oozie-hado-W
------------------------------------------------------------------------------------------------------------
Workflow Name : map-reduce-wf
App Path      : hdfs://hadoop:8020/user/hadoop/oozie/apps/map-reduce/workflow.xml
Status        : SUCCEEDED
Run           : 0
User          : hadoop
Group         : -
Created       : 2018-07-19 09:41 GMT
Started       : 2018-07-19 09:41 GMT
Last Modified : 2018-07-19 09:41 GMT
Ended         : 2018-07-19 09:41 GMT
CoordAction ID : -
```

```
Actions
------------------------------------------------------------------------------------------------
ID                                              Status    Ext ID    Ext Status    Err Code
------------------------------------------------------------------------------------------------
0000000-180717091145523-oozie-hado-W@:start:     OK        -         OK                    -
------------------------------------------------------------------------------------------------
0000000-180717091145523-oozie-hado-W@shell-node  OK    job_1531978002849_0028 SUCCEEDED -
------------------------------------------------------------------------------------------------
0000000-180717091145523-oozie-hado-W@end         OK        -         OK                    -
------------------------------------------------------------------------------------------------
```

⑨ 在 HDFS 文件系统上查看结果。

```
[hadoop@hadoop ~]$ hdfs dfs -cat /user/hadoop/oozie/output/search_out/part-r-00000
location1.txt    二把刀---|---光曜二---|---星曜二---|---
location2.txt    病态二公主---|---二当家---|---一个帅二个呆---|---
location3.txt    二营长---|---叫我二宝贝---|---二叉树---|---
```

9.3.5 技能实训

编写词频统计（WordCount）功能，使用 Oozie 工作流运行该程序。

实现步骤：

（1）编写 WordCount 程序。

（2）将程序打包提交到 Oozie 环境中运行。

本章总结

- Oozie 是一个工作流引擎服务器，其内部实现是一个 Java Web 程序。
- Oozie 以 Action 为基本单位，可以采用将多个 Action 构成一个 DAG 的模式运行。
- Oozie 通过 hPDL（一种通过 XML 自定义处理的语言）来构造工作流。

本章作业

一、简答题

1．简述 Oozie 三个组件的概念及关系。

2．如何定义一个工作流。

二、编码题

1．使用 Shell Action 在 HDFS 上创建/user/hadoop/student 目录。

2．使用 MapReduce Action 完成第 3 章本章作业编码题 1。

3．使用 MapReduce Action 完成第 3 章本章作业编码题 2。

第 10 章 项目实训——《王者荣耀》游戏英雄排行榜

技能目标

- 学会使用 HDFS 文件系统存储文件
- 学会集成 MapReduce 与 HBase
- 学会使用 HBase MapReduce API 操作 HBase 表
- 学会使用 Oozie 来管理任务调度

本章任务

任务：完成《王者荣耀》游戏英雄排行榜

本章资源下载

第10章 项目实训——《王者荣耀》游戏英雄排行榜

10.1 项目需求
10.2 项目环境准备
10.3 项目覆盖的技能点
10.4 难点分析
10.5 项目实现思路

10.1 项目需求

1. 背景

《王者荣耀》作为全球首款 5V5 英雄对战手游（手机游戏简称），是目前最热门的手游之一。很多游戏玩家都想知道《王者荣耀》游戏中，什么英雄比较热门，哪些英雄最受玩家喜爱。这就需要对英雄的使用情况进行统计，分析、计算出英雄的使用排行榜。

2. 需求

本项目要实现的功能是从某一时间段收集的游戏记录数据中统计出每个英雄的使用次数，然后对使用次数进行排序，得出《王者荣耀》游戏中英雄的排行榜。

项目的源数据如表 10-1 所示。

表 10-1 《王者荣耀》游戏记录数据

记录标识	英雄编号	英雄名称	英雄定位	玩家名称	游戏结果	游戏时间
1_001_2018-07-01	001	苏烈	坦克	你的仓鼠太爱我	1	2018-07-01
2_002_2018-07-01	002	刘邦	坦克	偷猪贼	1	2018-07-01
3_003_2018-07-01	003	钟馗	坦克	吹牛大王	0	2018-07-01
4_004_2018-07-02	004	张飞	坦克	双木夕	1	2018-07-02
5_005_2018-07-03	005	孙策	战士	浅念	1	2018-07-03
6_006_2018-07-04	006	哪吒	战士	星曜二	0	2018-07-04
7_007_2018-07-04	007	关羽	战士	光曜二	0	2018-07-04
8_008_2018-07-05	008	司马懿	法师	会飞的猪	1	2018-07-05
9_021_2018-07-05	021	杨玉环	法师	广东省第一蔡文姬	1	2018-07-05
10_003_2018-07-06	003	钟馗	坦克	国士无双	1	2018-07-05
…	…	…	…	…	…	…

其中，“游戏结果”列中，1 表示“胜利”，0 表示“失败”。表 10-1 中的数据只是文件中的一部分。

项目最终运行结果如图 10.1 所示。

```
[hadoop@hadoop ~]$ hdfs dfs -cat /user/hadoop/tophero/part-r-00000
6	小乔
6	韩信
4	钟无艳
3	武则天
3	明世隐
2	成吉思汗
2	钟馗
2	花木兰
2	百里玄策
2	孙膑
2	孙策
2	哪吒
2	司马懿
2	关羽
2	元歌
1	李元芳
1	杨玉环
1	张飞
1	苏烈
1	刘邦
1	公孙离
```

图10.1 《王者荣耀》游戏英雄排行榜

3. **分析**

本章要完成的《王者荣耀》游戏英雄排行榜需要做到：

- 使用 HDFS 保存游戏记录数据文件。
- 使用 HBase 的 MapReduce 应用工具 Bulk Loading 将数据批量导入到 HBase 表中。
- 使用 MapReduce 作业实现统计英雄使用次数并降序输出。
- 使用 HBase 数据库存储英雄排行榜记录。
- 使用 Oozie 完成上述步骤的作业调度。

10.2 项目环境准备

实现《王者荣耀》游戏英雄排行榜功能，对于开发环境的要求如下：

- 开发工具：IntelliJ IDEA，JDK1.8、apache-maven-3.5.3。
- 开发语言：Java、Shell。
- 软件工具及版本：

CentOS 7.3-1611
VMware12 pro
MySQL 5.6.40
Hadoop-2.6.0-cdh5.14.2
ZooKeeper-3.4.5-cdh5.14.2
HBase-1.2.0-cdh5.14.2
Oozie-4.1.0-cdh5.14.2

10.3 项目覆盖的技能点

项目覆盖的技能点如下：

- Shell 操作 HDFS。
- HDFS、MapReduce 与 HBase 的集成。
- 会使用 Bulk Loading 工具完成 HBase 快速批量数据导入。
- 利用 MapReduce 接收 HBase 数据输入。
- 利用 MapReduce 输出数据到 HBase。
- 利用 Oozie 完成作业调度。

10.4 难点分析

1. **搭建环境**

本项目采用的环境是在 VMware 创建的虚拟机上安装 Hadoop、ZooKeeper、HBase、Oozie 软件，在前面已经分别介绍过这 4 个软件的安装及配置，读者可以参照对应内容搭建环境，这里不再赘述。

除了上述的环境以外，还需要将 HBase 作为 MapReduce 作业的输入和输出，来完成数据的统计，所以要完成 MapReduce 与 HBase 的集成。使用 HBase 和 MapReduce 协同工作，既利用了 MapReduce 分布式计算的优势，也利用了 HDFS 的海量存储，特别是利用了 HBase 可以对海量数据实时访问的特点。

默认情况下，MapReduce 作业发布到集群后，不能访问 HBase 的配置文件和相关类，首先需要对集群中各节点的 Hadoop 环境做如下调整。

（1）将 hbase-site.xml 复制到 Hadoop 集群每个节点$HADOOP_HOME/etc/hadoop 目录下。

（2）编辑$HADOOP_HOME/etc/hadoop/hadoop-env.sh 文件，增加下面一行。

```
export HADOOP_CLASSPATH=$HADOOP_CLASSPATH:$HBASE_HOME/lib/*
```

其中，调整（1）使 MapReduce 作业在运行时可以连接到 ZooKeeper 集群；调整（2）将 HBase 安装目录 lib 下的所有 jar 文件添加到环境变量$HADOOP_CLASSPATH 中，使得 MapReduce 作业可以访问所依赖的 HBase 相关类，从而不必每次都将 HBase 相关类打包到 MapReduce 应用的 jar 文件中。最后注意，如果 Hadoop 使用的是完全分布式的环境，还要将上述两个文件复制到 Hadoop 集群中的其他节点上。使用如下命令可以测试环境是否正确配置。

```
hadoop jar $HBASE_HOME/lib/hbase-server-1.2.0-cdh5.14.2.jar rowcount gamer
```

该命令运行 hbase-server-1.2.0-cdh5.14.2.jar 中的 MapReduce 应用 rowcount，参数为 HBase 中的表名“gamer”，其功能是使用 MapReduce 框架统计 HBase 数据库表 gamer 中的行数。

2. **HBase 表的设计**

在 HBase 中，行键的设置对表的性能有很大的影响，根据表 10-1 的数据以及需求分析设计的 HBase 表如下。

- hero 表
- herolist 表

3. Bulk Loading 数据导入

使用 HBase 客户端 API 可以将《王者荣耀》游戏记录数据导入到 HBase 表中，但这种手动编写代码的方式效率过于低下。在 HBase 中，提供了 HBase MapReduce 应用——Bulk Loading 工具 importtsv 和 completebulkload，它们实质上是 MapReduce 作业。importtsv 能够按 HBase 内部数据格式输出表数据，completebulkload 能够直接加载生成的 HFile 到 HDFS，使用这两种工具比单纯使用 HBase 客户端 API 逐条写入数据库要节省 CPU 和网络资源。

使用 Bulk Loading 加载工具有两个步骤。

（1）通过 importtsv 准备数据，运行以下命令可以从 MapReduce 作业生成 HBase 数据文件 HFile 到指定的目录。

```
hadoop jar $HBASE_HOME/lib/hbase-server-1.2.0-cdh5.14.2.jar importtsv
        -D importtsv.bulk.output=tmp
        -D importtsv.columns=HBASE_ROW_KEY,family:qualifier tablename /input/
```

在上面的命令中，importtsv 使用“-D”设置了两个选项：

- -Dimporttsv.bulk.output=tmp 指定生成的 HFile 位置，位于 HDFS 用户根目录下的 tmp 目录中。
- -Dimporttsv.columns=HBASE_ROW_KEY,family:qualifier 对应输入文件中的行记录（读入的行默认以 Tab 键分割）和对应表的列格式，HBASE_ROW_KEY 表示行键，后续参数为列键。tablename 表示操作对应的表。最后的参数/input/表示输入文件目录（HDFS），命令运行成功后，在用户根目录下的 tmp 目录下可以找到生成的 HFile 文件。

（2）使用 completebulkload 将 HFile 移动到 HRegion 中完成数据加载。

hadoop jar $HBASE_HOME/lib/ hbase-server-1.2.0-cdh5.14.2.jar completebulkload tmp tablename

其中，tmp 为生成的 HFile 所在的目录，tablename 为操作的表。

上述操作过程中可能会存在的问题及解决方式如下。

- ClassNotFound：检查 MapReduce 和 HBase 集成环境是否配置正确。
- 表不存在：先创建表或者对 importtsv 使用“-Dcreate.table=yes”选项。
- tmp 目录已存在：在 HDFS 上删除 tmp 目录。
- ZooKeeper session 超时或过期：尝试重新启动 HBase。

4. 编写 MapReduce 程序

HBase Java API 对 MapReduce API 进行了扩展，称为 HBase MapReduce API。显然这是 HBase 提供的，主要是为了方便 MapReduce 应用对 HTable 的操作。要实现《王者荣耀》游戏英雄排行榜的统计及排序功能，需要使用 HBase MapReduce API 编写 MapReduce 程序。

前面提过，MapReduce 的输入/输出包括数据输入/输出的文件和处理这些文件所采用的输入/输出格式。与 HBase 集成后，“输入/输出的文件”变为“表（HTable）”，那么对表的输入/输出格式也要提供相应的实现，分别是 TableInputFormat 和 TableOutputFormat，它们所在的 jar 文件名为 hbase-server-1.2.0-cdh5.14.2.jar，该 jar 文件存放在$HBASE_HOME/lib 目录下。同时 HBase 还提供了 TableMapper 和 TableReducer 类，使得编写 MapReduce 程序更加方便。HBase MapReduce API 与 Hadoop MapReduce API 的关系如表 10-2 所示。

表 10-2 HBase MapReduce API 与 Hadoop MapReduce API 的关系

HBase MapReduce API	Hadoop MapReduce API
org.apache.hadoop.hbase.mapreduce.TableMapper	org.apache.hadoop.mapreduce.Mapper
org.apache.hadoop.hbase.mapreduce.TableReducer	org.apache.hadoop.mapreduce.Reducer
org.apache.hadoop.hbase.mapreduce.TableInputFormat	org.apache.hadoop.mapreduce.InputFormat
org.apache.hadoop.hbase.mapreduce.TableOutputFormat	org.apache.hadoop.mapreduce.OutputFormat

表 10-2 中左侧的类均继承于右侧的类，表中列出了 HBase MapReduce API 的主要类；TableMapper、TableReducer、TableInputFormat、TableOutputFormat，均位于 hbase-server-1.2.0-cdh5.14.2.jar 中。除此之外，集成 MapReduce 和 HBase 还需要两个 jar 文件。新增三个 jar 文件：hbase-server-1.2.0-cdh5.14.2.jar、hbase-hadoop-compact-1.2.0-cdh5.14.2.jar、metrics-core-2.2.0.jar，与之前的 HBase 所需的 jar 文件一起添加到 MapReduce 工程的构建路径当中。

与之前集成 Mapper 和 Reducer 相比，使用继承 TableMapper 和 TableReducer 的方式编写 MapReduce 作业来读写 HBase 数据库表会更方便。首先注意新旧 MapReduce API 有区别，在此一律使用 org.apache.hadoop.hbase.mapreduce 包下的相关类。下面分别从 Map 和 Reduce 两个方面来讲解 HBase MapReduce API。

（1）TableMapper

TableMapper 继承于 Mapper 类，其类定义如下。

```
public abstract class TableMapper<KEYOUT,VALUEOUT> extends
    Mapper<ImmutableByteWritable,Result,KEYOUT,VALUEOUT>{
}
```

可以看出 TableMapper 没有实现任何功能，只是一个不用提供 KEYIN 和 VALUEIN 参数的 Mapper。也就意味着，TableMapper 限制了每一行输入数据的键/值类型，分别为 ImmutableBytesWritable 和 Result，而输出的键/值类型由提供的 KEYOUT 和 VALUEOUT 两个参数来指定。

HBase 为 MapReduce 提供了专用输入格式 TableInputFormat，而 TableInputFormat 实现了 InputFormat<ImmutableBytesWritable,Result>接口，能够读取 HBase 行数据并封装为<ImmutableBytesWritable,Result>格式的键/值对，然后将其传递给 TableMapper，显然 TableMapper 获得的行数据总是该格式，也就不用再指定输入数据的键/值类型了。

Result 类型表示 HBase 中的一行数据，而 ImmutableBytesWritable 表示行键。ImmutableBytesWritable 是不可变的字节数组类型，它是对“byte[]”的封装，不同之处

只是在 ImmutableBytesWritable 中读写时其参数限制为 final。

（2）TableReducer

org.apache.hadoop.hbase.mapreduce.TableReducer 继承自 Reducer 类，是由 HBase 提供的专用于输出到 HBase 数据库表的 Reducer 类型。其类定义如下。

```
public abstract class TableReducer<KEYIN,VALUEIN,KEYOUT>
        extends Reducer<KEYIN,VALUEIN,KEYOUT,Mutation>{
}
```

同 TableMapper 类似，TableReducer 也仅仅对所需输入与输出的键/值类型形参做了相应定义。在使用 TableReducer 时需要指定三个参数。

- KEYIN：输入键类型，对应之前 map 任务的输出键类型。
- VALUEIN：输入值类型，对应之前 map 任务的输出值类型。
- KEYOUT：输出键类型，reduce 任务（自身）的输出键类型。

TableReducer 默认的输出值类型为 Mutation，Mutation 是 HBase 中 Put、Get、Delete、Append 的父类。与 TableOutputFormat 类的输出值类型保持一致，基本上这两个类也总是结合在一起使用，因此必须确保两者的输出键/值类型相同。

在《王者荣耀》游戏英雄排行榜中，reduce 任务和单词统计程序类似，map 任务的输出结果经过 shuffle 处理后作为 reduce 阶段的输入，此时 reduce 每行输入格式应为 <key,values[]>。在此基础上对每个键/值对进行分析归纳，最后生成新的键/值对再输出。

（3）运行时第三方 jar 的依赖添加

MapReduce 作业的运行过程可分为两个阶段：客户端执行阶段和任务运行阶段。客户端需要被打包成 jar 文件并复制到集群上，以“hadoop jar <打包 jar 文件> <客户端驱动类> <参数>”命令方式执行。由驱动类的 main()方法驱动作业运行，最后向 ResourceManager（YARN）提交作业，直到提交完成前，都是客户端执行阶段。从作业提交后开始初始化、执行任务直到任务结束，都是任务执行阶段。

① 客户端执行阶段的类路径

在 MapReduce 执行阶段，执行的是驱动类的 main()方法。由“hadoop jar <jar>”设置客户端的类路径。如果存在第三方依赖 jar 文件，要么将其放到项目 lib 目录下，要么将其添加到$HADOOP_CLASSPATH 中。

② 任务执行阶段的类路径

在 MapReduce 任务执行阶段，执行的是 Mapper、Reducer 的 map()和 reduce()方法。在集群上，任务在各自的 JVM 上运行，它们的类路径不受$HADOOP_CLASSPATH 控制，也就是说，$HADOOP_CLASSPATH 值只对客户端有效，运行时需要的第三方 jar 则需要单独处理。

本项目案例中，HBase 的库对于 Hadoop 任务来说，就属于第三方库，所以也需要在运行时进行处理。

为了解决 MapReduce 运行时第三方 jar 文件的依赖问题，Hadoop 提供了两个工具类：GenericOptionsParser 和 TableMapReduceUtil。

GenericOptionsParser 是 Hadoop 的辅助类，该类会读取相关命令行参数。GenericOptionsParser 的选项使用如表 10-3 所示。

表 10-3　GenericOptionsParser 的选项

选项	说明
-Dproperty=value	将指定值赋给某个 Hadoop 配置属性，如前面介绍的“-D importtsv.bulk.output=tmp”。用于覆盖配置文件里的默认属性或站点属性，或通过-conf 选项设置的任何属性
-conf filename ...	将指定的文件添加到配置的资源列表中，等价于 conf.addResource(filename)
-fs uri	用指定的 uri 设置默认文件系统
-libjars jar1,jar2,...	从本地文件系统（或任务指定的文件系统）中复制指定的 jar 文件到 HDFS，确保 MapReduce 可以访问这些文件
-archives archive1,archive2,...	从本地文件系统（或任务指定的文件系统）中复制指定的存档到 HDFS，确保 MapReduce 可以访问这些文档

TableMapReduceUtil 是对 HBase 中的表进行 MapReduce 编程的工具类，用于配置 TableMapper 和 TableReducer，该类的方法都是静态方法，如表 10-4 所示。

表 10-4　TableMapReduceUtil 的方法

返回值	方法名	描述
void	initTableMapperJob(String,Scan,Class<? extends TableMapper>, Class<?>,Class<?>,Job)	配置作业的 map 任务，参数分别表示待扫描的表名、列族、列、TableMapper 及其输出键/值类型和 Job 对象
void	initTableReducerJob(String,Class<? extends TableReducer>, Job)	配置作业的 reduce 任务，参数分别表示输出的表名、TableReducer 和 Job 对象
void	addDependencyJars(Configuration，Class<?>...)	为作业添加包含指定类的 jar 文件，保存至集群分布式缓存
void	addDependencyJars(Job)	为作业添加 HBase 依赖库，保存至集群分布式缓存

注意

使用 initTableMapperJob()方法后，作业的输入格式类型固定为 TableInputFormat。

使用 initTableReducerJob()方法后，作业的输出格式类型固定为 TableOutputFormat。

5. Oozie 作业调度

从前面的需求分析中可以看出，从导入数据到最后计算出结果并显示，每一个环节之间都存在一定的依赖，最后会采用 Oozie 来管理这些任务，实现作业的自动化调度执行。

执行流程如图 10.2 所示。

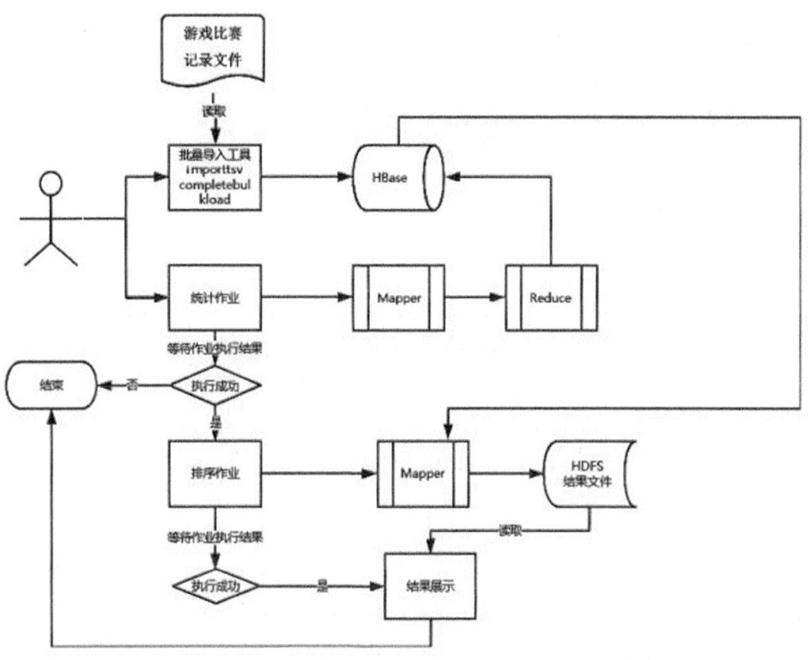

图10.2　执行流程图

10.5 项目实现思路

1. 数据文件存储

（1）在 HDFS 上创建存放《王者荣耀》游戏数据文件的目录。

```
[hadoop@hadoop ~]$ hdfs dfs -mkdir -p /input/bulkload
```

（2）上传《王者荣耀》游戏数据文件到 HDFS 的/input/bulkload 目录下。

```
[hadoop@hadoop ~]$ hdfs dfs -put ~/data/hero.txt    /input/bulkload
```

（3）查看是否创建成功。

```
[hadoop@hadoop ~]$ hdfs dfs -ls /input/bulkload
Found 1 items
-rw-r--r--      1 hadoop supergroup            2800 2018-07-22 01:45 /input/bulkload/hero.txt
```

2. HBase 表的创建

在 HBase 中创建数据表 hero 和 herolist。

（1）启动 HBase Shell。

```
[hadoop@hadoop ~]$ hbase shell
```

（2）在 HBase 中创建 hero 表。

```
hbase(main):001:0> create 'hero','info'
0 row(s) in 1.3690 seconds

     => Hbase::Table - hero
```

（3）在 HBase 中创建 herolist 表。

```
hbase(main):002:0> create 'herolist','details'
0 row(s) in 1.3330 seconds

    => Hbase::Table - herolist
```

3．数据批量导入

（1）将 HDFS 目录/input/bulkload 下的文件通过 importtsv 工具生成 HBase 数据文件。

```
[hadoop@hadoop ~]$ hadoop jar /opt/hbase-1.2.0-cdh5.14.2/lib/hbase-server-1.2.0-cdh5.14.2.jar \
importtsv -Dcreate.table=no –Dimporttsv.bulk.output=/user/hadoop/tmp –D \
importtsv.columns=HBASE_ROW_KEY,info:herocode,info:heroname,info:herodefinition,info:playername,
info:gameresult,info:gametime hero /input/bulkload
```

（2）执行完以后，可以查看“–Dimporttsv.bulk.output”指定的目录/user/hadoop/tmp 下是否存在 HFile 文件。

```
[hadoop@hadoop ~]$ hdfs dfs -ls /user/hadoop/tmp
Found 2 items
-rw-r--r--    1  hadoop supergroup          0 2018-07-22 22:59 /user/hadoop/tmp/_SUCCESS
drwxr-xr-x    -  hadoop supergroup          0 2018-07-22 22:59 /user/hadoop/tmp/info
```

（3）使用 completebulkload 将 HFile 移动到 HRegion 中，完成将数据加载到 Hero 表。

```
[hadoop@hadoop ~]$hadoop jar /opt/hbase-1.2.0-cdh5.14.2/lib/hbase-server-1.2.0-cdh5.14.2.jar
completebulkload tmp hero
```

（4）导入完成以后，可以通过 HBase Shell 查看表中数据。

```
hbase(main):001:0> scan 'hero'
ROW                          COLUMN+CELL
 10_003_2018-07-05           column=info:gameresult, timestamp=1532279398431, value=1
 10_003_2018-07-05           column=info:gametime, timestamp=1532279398431, value=2018-07-05
 10_003_2018-07-05           column=info:herocode, timestamp=1532279398431, value=003
 10_003_2018-07-05           column=info:herodefinition, timestamp=1532279398431,
                             value=\xE5\x9D\xA6\xE5\x85\x8B
 10_003_2018-07-05           column=info:heroname, timestamp=1532279398431,
                             value=\xE9\x92\x9F\xE9\xA6\x97
 10_003_2018-07-05           column=info:playername, timestamp=1532279398431,
                             value=\xE5\x9B\xBD\xE5\xA3\xAB\xE6\x97\xA0\xE5\x8F\x8C
 11_017_2018-07-05           column=info:gameresult, timestamp=1532279398431, value=1
 11_017_2018-07-05           column=info:gametime, timestamp=1532279398431, value=2018-07-05
 11_017_2018-07-05           column=info:herocode, timestamp=1532279398431, value=017
 11_017_2018-07-05           column=info:herodefinition, timestamp=1532279398431,
                             value=\xE6\x88\x98\xE5\xA3\xAB
 11_017_2018-07-05           column=info:heroname,timestamp=1532279398431,
                             value=\xE9\x9F\xA9\xE4\ xBF\xA1
 11_017_2018-07-05           column=info:playername, timestamp=1532279398431,
                             value=\xE4\xB8\x89\xE9\xA2\x97\xE6\x9F\x9A
 ...
```

4. **数据统计和排序功能**

编写 MapReduce 程序实现对《王者荣耀》游戏记录表中英雄的使用次数进行统计并进行降序显示。

实现步骤：

（1）使用 IntelliJ IDEA+Maven 创建项目工程及依赖。

（2）编写代码完成英雄使用次数的统计功能。

（3）编写代码完成对使用次数排序的功能。

（4）配置作业。

pom 文件添加的依赖包内容如下。

```
<dependency>
    <groupId>org.apache.hbase</groupId>
    <artifactId>hbase-client</artifactId>
    <version>1.2.0-cdh5.14.2</version>
</dependency>
<dependency>
    <groupId>org.apache.hbase</groupId>
    <artifactId>hbase-common</artifactId>
    <version>1.2.0-cdh5.14.2</version>
</dependency>
<dependency>
    <groupId>org.apache.hbase</groupId>
    <artifactId>hbase-protocol</artifactId>
    <version>1.2.0-cdh5.14.2</version>
</dependency>
<dependency>
    <groupId>org.apache.hbase</groupId>
    <artifactId>hbase-server</artifactId>
    <version>1.2.0-cdh5.14.2</version>
</dependency>
<dependency>
    <groupId>org.apache.hbase</groupId>
    <artifactId>hbase-hadoop-compat</artifactId>
    <version>1.2.0-cdh5.14.2</version>
</dependency>
<dependency>
    <groupId>org.apache.hadoop</groupId>
    <artifactId>hadoop-common</artifactId>
    <version>2.6.0-cdh5.14.2</version>
</dependency>
<dependency>
    <groupId>org.apache.hadoop</groupId>
    <artifactId>hadoop-hdfs</artifactId>
    <version>2.6.0-cdh5.14.2</version>
```

```
</dependency>
<dependency>
    <groupId>org.apache.hadoop</groupId>
    <artifactId>hadoop-mapreduce-client-core</artifactId>
    <version>2.6.0-cdh5.14.2</version>
</dependency>
<dependency>
    <groupId>org.apache.hadoop</groupId>
    <artifactId>hadoop-mapreduce-client-jobclient</artifactId>
    <version>2.6.0-cdh5.14.2</version>
    <scope>test</scope>
</dependency>
<dependency>
    <groupId>org.apache.hadoop</groupId>
    <artifactId>hadoop-mapreduce-client-common</artifactId>
    <version>2.6.0-cdh5.14.2</version>
</dependency>
<dependency>
    <groupId>commons-logging</groupId>
    <artifactId>commons-logging</artifactId>
    <version>1.2</version>
</dependency>
<dependency>
    <groupId>log4j</groupId>
    <artifactId>log4j</artifactId>
    <version>1.2.17</version>
</dependency>
<dependency>
    <groupId>junit</groupId>
    <artifactId>junit</artifactId>
    <version>4.11</version>
</dependency>
<dependency>
    <groupId>com.google.guava</groupId>
    <artifactId>guava</artifactId>
    <version>15.0</version>
</dependency>
<dependency>
    <groupId>com.yammer.metrics</groupId>
    <artifactId>metrics-core</artifactId>
    <version>2.2.0</version>
</dependency>
<dependency>
    <groupId>commons-collections</groupId>
    <artifactId>commons-collections</artifactId>
```

```
        <version>3.2.2</version>
    </dependency>
    <dependency>
        <groupId>org.slf4j</groupId>
        <artifactId>slf4j-api</artifactId>
        <version>1.7.25</version>
    </dependency>
    <dependency>
        <groupId>org.slf4j</groupId>
        <artifactId>slf4j-log4j12</artifactId>
        <version>1.7.25</version>
    </dependency>
    <dependency>
        <groupId>log4j</groupId>
        <artifactId>log4j</artifactId>
        <version>1.2.17</version>
    </dependency>
    <dependency>
        <groupId>commons-configuration</groupId>
        <artifactId>commons-configuration</artifactId>
        <version>1.10</version>
    </dependency>
    <dependency>
        <groupId>org.apache.hadoop</groupId>
        <artifactId>hadoop-auth</artifactId>
        <version>3.1.0</version>
    </dependency>
    <dependency>
        <groupId>commons-cli</groupId>
        <artifactId>commons-cli</artifactId>
        <version>1.2</version>
    </dependency>
    <dependency>
        <groupId>com.google.protobuf</groupId>
        <artifactId>protobuf-java</artifactId>
        <version>2.5.0</version>
    </dependency>
    <dependency>
        <groupId>org.apache.avro</groupId>
        <artifactId>avro</artifactId>
        <version>1.7.6-cdh5.14.2</version>
    </dependency>
    <dependency>
        <groupId>org.cloudera.htrace</groupId>
        <artifactId>htrace-core</artifactId>
```

```
        <version>2.04</version>
    </dependency>
    <dependency>
        <groupId>org.apache.htrace</groupId>
        <artifactId>htrace-core</artifactId>
        <version>3.1.0-incubating</version>
    </dependency>
    <dependency>
        <groupId>org.apache.htrace</groupId>
        <artifactId>htrace-core4</artifactId>
        <version>4.0.1-incubating</version>
    </dependency>
    <dependency>
        <groupId>io.netty</groupId>
        <artifactId>netty-all</artifactId>
        <version>4.1.25.Final</version>
    </dependency>
```

关键代码：

```
import java.io.InputStream;
import java.util.List;
import org.apache.hadoop.fs.FileSystem;
import org.apache.hadoop.hbase.Cell;
import org.apache.hadoop.hbase.CellUtil;
import org.apache.hadoop.hbase.HBaseConfiguration;
import org.apache.hadoop.hbase.client.Put;
import org.apache.hadoop.hbase.client.Scan;
import org.apache.hadoop.hbase.mapreduce.TableMapReduceUtil;
import org.apache.hadoop.hbase.mapreduce.TableMapper;
import org.apache.hadoop.hbase.mapreduce.TableReducer;
import org.apache.hadoop.hbase.util.Bytes;
import org.apache.hadoop.io.IOUtils;
import org.apache.hadoop.io.IntWritable;
import org.apache.hadoop.io.Text;
import org.apache.hadoop.hbase.client.Result;
import org.apache.hadoop.hbase.io.ImmutableBytesWritable;
import org.apache.hadoop.io.WritableComparable;
import org.apache.hadoop.mapreduce.lib.output.FileOutputFormat;
import org.apache.hadoop.mapreduce.Job;
import java.io.IOException;
import org.apache.hadoop.conf.Configuration;
import org.apache.hadoop.fs.Path;
import org.apache.hadoop.util.GenericOptionsParser;
public class TopHero {
    static final String TABLE_GAMERS="hero";
    static final String TABLE_NAMELIST="herolist";
```

```
static final String OUTPUT_PATH="tophero";
/**
 * 统计功能的 Map 方法
 */
static class GamersMapper extends TableMapper<Text,IntWritable> {
    protected void map(ImmutableBytesWritable key, Result value, Context context) throws
     InterruptedException, IOException {
        List<Cell> cells = value.listCells();
        for (Cell cell : cells) {
            if (Bytes.toString(CellUtil.cloneFamily(cell)).equals("info") &&
                    Bytes.toString(CellUtil.cloneQualifier(cell)).equals("heroname")) {
                context.write(new Text(Bytes.toString(CellUtil.cloneValue(cell))),
                            new IntWritable(1));
            }
        }
    }
}

/**
 * 统计功能的 Reduce 方法
 */
static class IntNumReducer extends TableReducer<Text,IntWritable,Text>{
    protected void reduce(Text key,Iterable<IntWritable> values,Context context)
      throws IOException, InterruptedException {
        int playCount = 0;
        for(IntWritable num:values){
            playCount+=num.get();
        }
        Put put = new Put(Bytes.toBytes(key.toString()));
        put.addColumn(Bytes.toBytes("details"),Bytes.toBytes("rank"),
          Bytes.toBytes(playCount));
        context.write(key,put);
    }
}
/**
 * 完成排序功能的 Map 方法
 */
static class GamersNameMapper extends TableMapper<IntWritable,Text>{
    protected void map(ImmutableBytesWritable key, Result value, Context
      context) throws InterruptedException, IOException {
        List<Cell> cells = value.listCells();
        for(Cell cell: cells){
                context.write(new IntWritable(
                        Bytes.toInt(CellUtil.cloneValue(cell))),
                        new Text(Bytes.toString(key.get())));
```

```
            }
        }
}
/**
 * 实现降序排序
 */
private static class IntWritableDecreaseingComparator extends
 IntWritable.Comparator{
     @Override
     public int compare(WritableComparable a, WritableComparable b) {
          return -super.compare(a, b);
     }
     @Override
     public int compare(byte[] b1, int s1, int l1, byte[] b2, int s2, int l2) {
          return -super.compare(b1, s1, l1, b2, s2, l2);
     }
}
/**
 * 配置作业：英雄使用次数统计
 */
static boolean gamerCount(String[] args) throws IOException,
 ClassNotFoundException, InterruptedException {
     Job job = Job.getInstance(conf);
     job.setJarByClass(TopHero.class);
     job.setNumReduceTasks(2);
     Scan scan = new Scan();
     TableMapReduceUtil.initTableMapperJob(
        TABLE_GAMERS,scan,GamersMapper.class,
        Text.class,IntWritable.class,job,false);
     TableMapReduceUtil.initTableReducerJob(
        TABLE_NAMELIST,IntNumReducer.class,job);
     return job.waitForCompletion(true);
}
/**
 * 配置作业排序
 */
static boolean sortGamers(String[] args) throws IOException,
 ClassNotFoundException, InterruptedException {
     Job job = Job.getInstance(conf);
     job.setJarByClass(TopHero.class);
     job.setNumReduceTasks(1);
     job.setSortComparatorClass(IntWritableDecreaseingComparator.class);
     TableMapReduceUtil.initTableMapperJob(TABLE_NAMELIST,new Scan(),
                              Gamers NameMapper.class,IntWritable.class,Text.class,job);
     Path output = new Path(OUTPUT_PATH);
```

```
        if(FileSystem.get(conf).exists(output)) {
            FileSystem.get(conf).delete(output, true);
        }
        FileOutputFormat.setOutputPath(job,output);
        return job.waitForCompletion(true);
    }
    /**
     * 查看输出文件
     * @throws IllegalArgumentException
     * @throws IOException
     */
    static void showResult() throws IllegalArgumentException,IOException {
        FileSystem fs =   FileSystem.get(conf);
        InputStream in = null;
        try {
            in = fs.open(new Path(OUTPUT_PATH+"/part-r-00000"));
            IOUtils.copyBytes(in,System.out,4096,false);
        }finally {
            IOUtils.closeStream(in);
        }
    }
    static Configuration conf = HBaseConfiguration.create();
    public static void main(String[] args) throws
     IOException,ClassNotFoundException,InterruptedException{
        GenericOptionsParser gop   = new GenericOptionsParser(conf,args);
        String[] otherArgs = gop.getRemainingArgs();
        if(gamerCount(otherArgs)){
            if(sortGamers(otherArgs)){
                showResult();
            }
        }
    }
}
```

5. Oozie 实现作业调度

使用 Oozie 对数据批量导入功能和数据统计及排序功能完成自动调度。

（1）在 HDFS 上创建 oozie 执行目录。

```
[hadoop@hadoop ~]$ hdfs dfs -mkdir -p /home/hadoop/oozie/apps/shell/bulkload
```

（2）创建本地的节点文件夹。

```
[hadoop@hadoop ~]$ mkdir -p /home/hadoop/oozie/bulkLoad
```

（3）封装 importtsv 命令。新建 hbase-script_01.sh 文件，并添加如下内容。

```
#!/bin/bash
hadoop fs -test -e /user/hadoop/tmp
if [ $? -eq 0 ];then
     hadoop fs -rm -r /user/hadoop/tmp
```

```
    if [ $? -eq 0 ];then
    hadoop jar /opt/hbase-1.2.0-cdh5.14.2/lib/hbase-server-1.2.0-cdh5.14.2.jar \
    importtsv -Dcreate.table=no -Dimporttsv.bulk.output=/user/hadoop/tmp \
    -Dimporttsv.columns=HBASE_ROW_KEY,info:herocode,info:heroname, \
    info:herodefinition,info:playername,info:gameresult,info:gametime    \
    hero /input/bulkload
    fi
else
    hadoop jar /opt/hbase-1.2.0-cdh5.14.2/lib/hbase-server-1.2.0-cdh5.14.2.jar importtsv \
    -Dcreate.table=no -Dimporttsv. bulk.output=/user/hadoop/tmp \
    -Dimporttsv.columns=HBASE_ROW_KEY,info:herocode,info:heroname,\
    info:herodefinition,info:playername,info:gameresult,info:gametime \
    hero /input/bulkload
fi
```

注意

这里为了让作业可以多次调用，使用了 shell 命令行来判断/user/hadoop/tmp 目录是否存在，如果存在，必须先删除，否则，import 命令会报错。

（4）封装 completebulkload 命令。新建 hbase-script_02.sh 文件，并添加如下内容。

```
#!/bin/bash
hadoop jar /opt/hbase-1.2.0-cdh5.14.2/lib/hbase-server-1.2.0-cdh5.14.2.jar \
completebulkload tmp hero
```

（5）在本地/home/hadoop/oozie/bulkLoad 目录下新建 lib 目录并将项目打成 jar 包上传到该目录下。

创建 lib 目录命令：

```
[hadoop@hadoop ~]$ mkdir –p /home/hadoop/oozie/bulkLoad/lib
```

注意

项目在集群中运行，因此需要添加 hbase-hadoop-compat-1.2.0-cdh5.14.2.jar 依赖包，读者需要下载该依赖包并添加到/home/hadoop/oozie/bulkLoad/lib 目录中。

（6）在/home/hadoop/oozie/bulkLoad/目录下新建 job.properties 文件，并添加如下内容。

```
nameNode=hdfs://hadoop:8020
jobTracker=hadoop:8032
queueName=default
examplesRoot=oozie
oozie.wf.application.path=
            ${nameNode}/home/${user.name}/${examplesRoot}/apps/shell/bulkload/workflow.xml
EXEC=hbase-script_01.sh
EXEC_02=hbase-script_02.sh
```

（7）在/home/hadoop/oozie/bulkLoad/目录下新建 workflow.xml 文件，并添加如下内容。

```
<workflow-app xmlns="uri:oozie:workflow:0.4" name="shell-wf">
        <start to="shell-node"/>
        <action name="shell-node">
            <shell xmlns="uri:oozie:shell-action:0.2">
                <job-tracker>${jobTracker}</job-tracker>
                <name-node>${nameNode}</name-node>
                <configuration>
                    <property>
                        <name>mapred.job.queue.name</name>
                        <value>${queueName}</value>
                    </property>
                </configuration>
                <exec>${EXEC}</exec>
                <file>${EXEC}#${EXEC}</file>
            </shell>
            <ok to="second"/>
            <error to="fail"/>
        </action>
        <action name="second">
            <shell xmlns="uri:oozie:shell-action:0.2">
                <job-tracker>${jobTracker}</job-tracker>
                <name-node>${nameNode}</name-node>
                <configuration>
                    <property>
                        <name>mapred.job.queue.name</name>
                        <value>${queueName}</value>
                    </property>
                </configuration>
                <exec>${EXEC_02}</exec>
                <file>${EXEC_02}#${EXEC_02}</file>
            </shell>
            <ok to="third"/>
            <error to="fail"/>
        </action>
        <action name="third">
            <java>
                    <job-tracker>${jobTracker}</job-tracker>
                    <name-node>${nameNode}</name-node>
                    <configuration>
                            <property>
                                    <name>mapred.job.queue.name</name>
                                    <value>${queueName}</value>
                            </property>
                    </configuration>
```

```
                <main-class>TopHero</main-class>
            </java>
            <ok to="end"/>
            <error to="fail"/>
        </action>
        <kill name="fail">
            <message>
                Shell action failed, error message[${wf:errorMessage(wf:lastErrorNode())}]
            </message>
        </kill>
        <end name="end"/>
    </workflow-app>
```

在 workflow.xml 中，将实现数据导入的两个命令操作和实现数据统计和排序功能的 MapReduce 功能作为三个 action，并且三个 action 之间是一个顺序执行的关系，一个动作节点执行成功，才能执行下一个动作节点。

（8）将 hbase-script_01.sh、hbase-script_02.sh、lib、job.properties、workflow.xml 上传到 HDFS 的/home/hadoop/oozie/apps/shell/bulkload 目录下。

```
[hadoop@hadoop ~]$ hdfs dfs -put /home/hadoop/oozie/bulkLoad/* /home/hadoop/oozie/apps/shell/bulkload
```

（9）运行 Oozie 调度任务。

```
[hadoop@hadoop ~]$ oozie job -oozie http://hadoop:11000/oozie -config /home/hadoop/oozie/bulkLoad/job.properties -run
job: 0000001-180722175151199-oozie-hado-W
```

（10）查看作业运行结果。

```
[hadoop@hadoop ~]$ oozie job -oozie http://hadoop:11000/oozie -info 0000001-180722175151199-oozie-hado-W
Job ID : 0000001-180722175151199-oozie-hado-W
------------------------------------------------------------------------------------------------------------
Workflow Name : shell-wf
App Path      : hdfs://hadoop:8020/home/hadoop/oozie/apps/shell/bulkload/workflow.xml
Status        : SUCCEEDED
Run           : 0
User          : hadoop
Group         : -
Created       : 2018-07-22 10:11 GMT
Started       : 2018-07-22 10:11 GMT
Last Modified : 2018-07-22 10:15 GMT
Ended         : 2018-07-22 10:15 GMT
CoordAction ID : -
Actions
------------------------------------------------------------------------------------------------------------
ID                                          Status    Ext ID    Ext Status    Err Code
------------------------------------------------------------------------------------------------------------
```

```
0000001-180722175151199-oozie-hado-W@:start:  OK     -                   OK                 -
-----------------------------------------------------------------------------------------------
0000001-180722175151199-oozie-hado-W@shell-node OK  job_1532250719547_0008 SUCCEEDED   -
-----------------------------------------------------------------------------------------------
0000001-180722175151199-oozie-hado-W@second OK   job_1532250719547_0010 SUCCEEDED  -
-----------------------------------------------------------------------------------------------
0000001-180722175151199-oozie-hado-W@third   OK  job_1532250719547_0011 SUCCEEDED  -
-----------------------------------------------------------------------------------------------
0000001-180722175151199-oozie-hado-W@end     OK     -                   OK                 -
-----------------------------------------------------------------------------------------------
```

从运行结果中可以看出，所有的作业都已经成功执行。

（11）查看结果文件。

```
[hadoop@hadoop ~]$ hdfs dfs -cat /user/hadoop/tophero/part-r-00000
5        韩信
4        小乔
4        钟无艳
3        武则天
3        明世隐
2        成吉思汗
2        钟馗
2        花木兰
2        百里玄策
2        孙膑
2        孙策
2        哪吒
2        司马懿
2        关羽
2        元歌
1        李元芳
1        杨玉环
1        张飞
1        苏烈
1        刘邦
1        公孙离
```

项目案例整个执行过程请扫描二维码。

项目案例执行过程

6. Oozie 定时作业配置

在 Oozie 架构中，可以使用 Coordinator 配置定时调度，实现任务在指定的时间执行。要实现一个 Oozie 循环的定时调度，需要三个文件。

- job.properties：记录 job 的属性。
- workflow.xml：使用 hPDL 定义任务的流程和分支。
- coordinator.xml：定义 workflow 的触发条件，实现定时触发，也可以重组多个 workflow 的运行顺序。

下面针对本章的案例来配置定时执行。

实现步骤：

（1）配置 job.properties 文件。

（2）配置 workflow.xml 文件。

（3）配置 coordinator.xml 文件。

关键代码：

（1）将 job.properties 文件的内容修改为如下。

```
nameNode=hdfs://hadoop:8020
jobTracker=hadoop:8032
queueName=default
examplesRoot=oozie
EXEC=hbase-script_01.sh
EXEC_02=hbase-script_02.sh# oozie
#coordinator.xml 在 hdfs 上的路径
oozie.coord.application.path=${nameNode}/home/${user.name}/${examplesRoot}/apps/shell/bulkload/
coordinator.xml
#workflow.xml 在 hdfs 上的路径
workflowAppUri=${nameNode}/home/${user.name}/${examplesRoot}/apps/shell/bulkload/workflow.
xml
#workflow 的名字
workflowName=workflow
#定时任务的开始时间，以 GMT 时区为准
start=2016-11-03T09:00+0800
#定时任务的结束时间
end=2019-07-30T16:00+0800
```

（2）workflow.xml 文件不需要调整，还是使用前面的文件及内容。

（3）新建 coordinator.xml 文件，并添加如下内容。

```
<coordinator-app name="coordinator" frequency="${coord:minutes(10)}" start="${start}" end="${end}"
timezone="GMT+0800" xmlns="uri:oozie:coordinator:0.2">
  <action>
    <workflow>
      <app-path>${workflowAppUri}</app-path>
      <configuration>
        <property>
          <name>jobTracker</name>
          <value>${jobTracker}</value>
        </property>
        <property>
          <name>nameNode</name>
          <value>${nameNode}</value>
        </property>
        <property>
          <name>queueName</name>
```

```
        <value>${queueName}</value>
      </property>
    </configuration>
  </workflow>
</action>
</coordinator-app>
```

其中，#frequency 的执行频率 ${coord:minutes(10)} 为 Coordinator 内置的 EL（Expression Language，表达式语言）函数。有关 Coordinator 内置 EL 函数的更多介绍请扫描二维码。

coordinator 内置的 EL 函数

本章总结

通过完成《王者荣耀》游戏英雄排行榜，读者进一步加深了对 HDFS、MapReduce、HBase、Oozie 的理解，能够熟练操作 HDFS、编写 MapReduce 程序、操作 HBase 以及进行 Oozie 作业调度。

动手能力是学习技术的终极目标。只有不断练习，不断写代码，不断完成更多的项目，才是提高动手能力的唯一途径。

本章作业

独立完成《王者荣耀》游戏英雄排行榜功能。